Metal Building Contracting and Construction

Metal Building Contracting and Construction

William D. Booth

McGraw-Hill
New York San Francisco Washington, D.C. Auckland Bogotá
Caracas Lisbon London Madrid Mexico City Milan
Montreal New Delhi San Juan Singapore
Sydney Tokyo Toronto

Library of Congress Cataloging-in-Publication Data

Booth, W. D.
Metal building contracting and construction / William D. Booth.
p. cm.
Includes index.
ISBN 0-07-006964-6
1. Building, Iron and steel. 2. Construction contracts.
3. Letting of contracts. 4. Small business—Management. I. Title.
TH1611.B676 1999
693'.7—dc21 98-43601
CIP

McGraw-Hill

A Division of The ***McGraw-Hill*** *Companies*

1 2 3 4 5 6 7 8 9 0 DOC/DOC 9 0 4 3 2 1 0 9

ISBN 0-07-006964-6

The sponsoring editor for this book was Larry Hager, the editing supervisor was Frank Kotowski, Jr., and the production supervisor was Pamela A. Pelton. It was set in Palatino per the BSF design by Kim Sheran and Paul Scozzari of McGraw-Hill's Professional Group Composition Unit in Hightstown, N.J.

Printed and bound by R. R. Donnelley & Sons.

McGraw-Hill books are available at special quantity discounts to use as premiums and sales promotions, or for use in corporate training programs. For more information, please write to the Director of Special Sales, McGraw-Hill, 11 West 19th Street, New York, NY 10011. Or contact your local bookstore.

This book is printed on recycled, acid-free paper containing a minimum of 50% recycled, de-inked fiber.

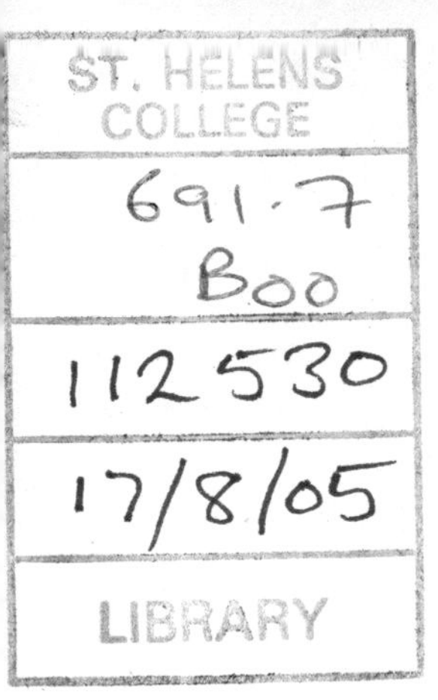

Acknowledgments

I would like to thank the below noted organizations for allowing their very important metal building information to be included in *Metal Building Contracting and Construction:*

American Institute of Steel Construction, Inc.

Building Officials and Code Administrators International, Inc.

Metal Building Manufacturers Association

The "Low Rise Market Share" figure on p. 1, the "New Construction Rises" figure on p. 2, the MBMA End-Use Report on pp. 3–5, Appendix A, and Appendix E* are courtesy of the Metal Building Manufacturers Association.

Appendix B, The BOCA National Building Code/1996: Copyright 1996, Building Officials and Code Administrators International, Inc., Country Club Hills, Illinois. *1996 BOCA Building Code*. Reprinted with permission of author. All rights reserved.

Appendix C, the brochure "Certified Excellence," is reproduced courtesy of the American Institute of Steel Construction, Inc., and the Metal Building Manufacturers Association.

Appendix D is a reprint of the article "Metal Building Systems: The Low-Rise Solution," courtesy of *Modern Steel Construction* (a publication of the American Institute of Steel Construction).

*From the AIA/*Architectural Record* Continuing Edication Series.

Contents

Introduction

With a jolt you pull your car to the side of the road and look at the metal building under construction. Then you note the job sign: New Home of ABC Inc.; General Contractor XYZ Construction Co.

You think out loud, "I knew the property had been sold," and your eyes roam over the real estate sign with the angled "Sold" sticker proclaiming that the agent has been successful. "Thought the job would have come out on the bid list by now, and why is a metal building being used?," you mutter as you move back onto the road, and head for your office.

Later in the day you get the real estate agent on the phone: "Jack how did XYZ get the ABC job? I saw nothing on the builders exchange about it, and what's really got me baffled is why is a metal building being used?"

"Bill, XYZ negotiated with ABC and put everything together under a design-build contract. They did one fine job of selling the owner. Understand they just called on him out of the blue asking for a chance, and from what I have heard it was the metal building that really caught the attention of ABC owner."

"Didn't he get another price Jack?"

"Sure did, two as a matter of fact. I put him on to two other contractors that could do metal buildings. He liked XYZ better. I don't know the prices, but I really feel those boys impressed him by going to him and making him aware of how the metal building would serve his needs now and in the future for a reasonable construction cost. Were you planning to bid the job?"

"Thinking about it, but then I don't do metal buildings either," you answer, trying to sound very casual.

"Well Bill maybe you should consider taking on a metal building line and the next time this situation comes up I'll recommend you."

"You know Jack I think you are right and I am going to look into this, but in the meantime how about 18 holes this Saturday at the club?"

This or something similar may have happened to you, and that's why you are looking into metal building contracting, or if you already are in the metal building field you are interested in increasing the business.

The purpose of this book is to give guidelines and answers to questions on how to approach and handle metal building contracting from the design-build direction. While bid work is geared to how inexpensively you can do the job, design-build lets you approach the owner and sell your services, including saving money and time, while maybe helping the customer get the jump on the competition. You sell the job, not bid it. I'm not saying you won't have any competition, as other construction companies may be after the same contract. You must convince the prospect that out of all the people he is talking with, you are the one who can offer the best deal, and quite often this does not mean the lowest price. A personal by-product of marketing construction is the excitement and fun of putting a deal together and making it work.

Every book on selling I've ever read does nothing but state broad principles about salesmanship. The small but pertinent questions concerning the details are never answered. As a matter of fact, they aren't even asked. In this book small details are the norm, not the exception.

This book does not give financial figures. It focuses on selling the job. Working knowledge of the economics of a construction project is a must, and I assume you know your way around in that area.

There are no guarantees! When all is said and done, it will still be up to you to make things happen.

William D. Booth

Chapter One

Figure 1-1. Metal buildings have come a long way in 50 years.

1

Metal Building Contracting

Metal Building Marketplace

Metal building contracting has become a very important section of the commercial construction industry. It fits in nicely with the firms that build for the business community.

The use of metal buildings has been on a steady rise for the past several years. Please note the Metal Building Manufacturers Association (MBMA) *Low-Rise Market* and the *New Construction* data sheets.

These two business reviews show that metal buildings are a large part of the commercial construction industry.

As noted in the Metal Building Manufacturers Association End-Use Report, *manufacturing* is the leading area with 46.33 percent. *Commercial* is second with 31.42 percent, and *community* is third with 13.66 percent. These three areas make up 91.41 percent of the metal building sales and offer a very good area in which to operate to the metal building contractor.

Low Rise Market Share Remains At All-Time High 69%

Metal building systems construction remained the overwhelming choice of construction professionals in 1997, repeating the 69% share of all sales in the low rise, non-residential market. This key segment consists of one- and two-story community, commercial, and industrial buildings of up to 150,000 sq. ft.

New Construction Rises To 388 Million Sq. Ft.

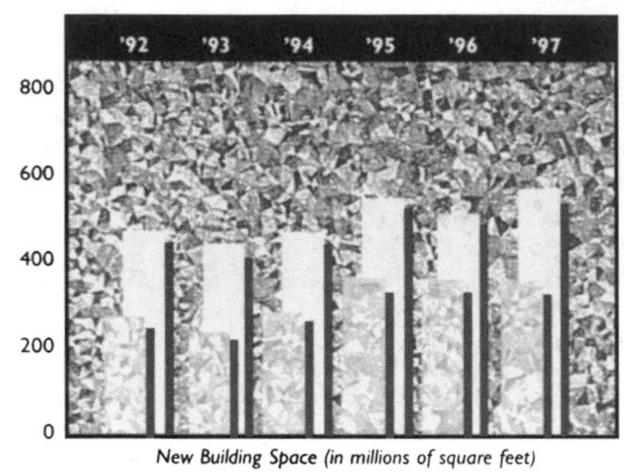

= *MBMA Sales* = *Industry sales*

During 1997, MBMA member companies accounted for 388 million out of 580 million sq. ft. of new building space completed in the low rise, nonresidential construction market. In addition, industrial buildings over 150,000 sq. ft. added 41.3 million sq. ft. of new space to overall MBMA sales.

Metal Building Partnership

The metal building contractor is somewhat like a car dealer. The car dealer will tie up with a car maker and sell that brand name exclusively. This allows the dealer and car manufacturer to form a partnership to sell and work together. The contractor who takes on a metal building line does the same thing. The building's brand name and the contractor's company name become a partnership; and when the metal building brand name is mentioned locally, the contractor's name will come up, because the business community has put the two together.

This partnership makes all the marketing assets of the manufacturer available to the contractor, so these assets can be put to good use to make money for the partnership.

Now, if you already have teamed up with a manufacturer, I'm sure you are well aware of what we are talking about; but if not, and if you are considering a metal building line, *the first place to start is the Metal Building Manufacturers Association* list of building manufacturers.

I recommend that you work with a metal building manufacturer who is a member of the MBMA because the association has standards that the members adhere to and work with. Please see members list on page 223 in App. A. Also, I think you will find very interesting the Metal Building Manufacturers Association and the American Institute of Steel Construction, Inc., information on *certified excellence* in App. C.

The metal building marketplace goes from the small building to the absolutely huge structure (see Figs. 2*a* and *b*), and this gives you, the contractor, a very important topic to consider: the size you are going to pursue. In my experience, most metal building contractors specialize in a

Metal Building Manufacturers Association End-Use Report, 1995–1997

	1995		1996		1997	
	No. of bldgs.	Dollars	No. of bldgs.	Dollars	No. of bldgs.	Dollars
1. Commercial storage for agricultural commodities	224	0.64%	334	0.97%	318	0.83%
2. Farm						
A. On farm commodity storage	227	0.50	242	0.38	192	0.39
B. Other farm buildings	1,222	1.47	1,122	1.47	947	1.04
	1,449	1.97%	1,364	1.85%	1,139	1.43%
3. Manufacturing						
A. Production	7,992	28.26	8,744	30.1	22,256	35.73
B. Warehousing	3,043	9.12	2,519	8.48	6,504	8.94
C. Equipment service	2,695	2.47	1,492	2.26	1,933	1.65
	13,730	39.84%	12,755	40.84%	30,693	46.33%
4. Commercial						
A. Retail stores	3,861	8.87	3,561	7.87	11,070	7.94
B. Warehousing and storage	9,375	19.79	7,583	19.27	10,183	16.17
C. Hangers	276	0.94	259	1.16	1,939	1.37
D. Freight terminals	380	1.18	546	1.84	395	0.99
E. Offices and banks	1,410	3.01	1,129	3.07	1,176	2.96

Metal Building Manufacturers Association End-Use Report, 1995–1997 (*Continued*)

	1995		1996		1997	
	No. of bldgs.	Dollars	No. of bldgs.	Dollars	No. of bldgs.	Dollars
F. Commercial garages and service stations (Auto.)	1,259	1.86	1,271	2.07	1,221	1.99
	16,561	35.64%	14,349	35.27%	25,984	31.42%
5. Community (public and private)						
A. Recreational	1,012	3.12	1,103	3.41	2,488	3.04
B. Educational	987	3.14	1,139	3.55	1,701	4.51
C. Hospital and health treatment	172	0.30	192	0.44	216	0.57
D. Houses of worship	683	1.60	559	1.51	1,136	2.21
E. Government administration and service	1,223	3.12	1,185	2.87	1,344	2.85
F. Passenger terminals	42	0.05	47	0.08	39	0.09
G. Residential	236	0.16	171	0.26	135	0.31
H. Correctional facilities	44	0.23	31	0.20	16	0.08
	4,399	11.72%	4,427	12.32%	7,075	13.66%
6. Government export	54	0.12%	72	0.44%	86	0.25%

7. All other			
A. Component parts	4.55	2.04	2.46
B. Retrofit roof systems	0.19	0.72	0.97
C. Retrofit wall systems	0.07	0.02	0.03
D. Other	5.26	5.53	2.62
	10.07%	8.31%	6.08%
Major end uses included in 7D	Mini storage Mining utilities	Roof and walls on new construction Barracks Work shop Garage (ind) Mini storage Funeral home	Roof and wall for new construction

SOURCE: Compiled by Thomas Associates, Inc.

building size that goes along, and fits in, with the size of their construction business. In other words, larger contractors market the larger metal buildings, say, 25,000 square feet (ft^2) and up, while the smaller boys operate on the smaller jobs, say 2000 to 15,000 ft^2, and every now and then both will get in the middle area.

Figure 1-2a. A small metal building.

Figure 1-2b. A large metal building.

Now let me mention this on the size you work with. If you are going to make metal buildings the main area of your business—that is, you plan to specialize in metal buildings—then I suggest you look into doing any size building.

Consider the entire metal building market from small to large and go where the money is; then the maintenance and repair of metal buildings can be a very profitable part of your business.

I present a very interesting source of metal building information from the magazine titled *Modern Steel Construction* in App. D. This article was written by one of the Metal Building Manufacturers Association officials and is a very good source of metal building information.

Let me bring up here what I feel is a very important subject. And that is, it is unwise to consider not teaming up with a building manufacturer. You may think, why go to all the trouble and team up, when you can simply shop around for the best metal building price and use that. First, many of the manufacturers will sell only to a dealer; second, you cannot take advantage of the many selling assets of the manufacturer unless you are a partner. These assets will be covered in detail as you get into the book. Also, don't be surprised if the manufacturer charges you a fee to hook up with its firm. This tells the manufacturer that you are really serious about marketing metal buildings.

Let me tell you about a very nice sideline that comes from the contractor network when you team up with a metal building manufacturer. You join a club that gives you contacts in the whole area in which the manufacturer sells. Your network contacts can cover large areas in a great many states, even the whole United States.

For whatever the reason, you may need to talk to someone who is in a certain place about something that concerns you and that certain place. So look up your network dealer in that place and contact her or him.

I have two good examples, one on the calling side and one on the call-receiving side.

My father-in-law was going to close down his old home and move to another location. He lived 350 miles (mi) away, so it was something we could not help with very easily. He needed a couple of people with a truck that he could direct back and forth over the weekend to the new place, which was not that far away, not some moving company.

He called us and said he was having trouble finding a couple of people to do the lifting and moving, and time was starting to be a problem. So the next day I called my network dealer in the town and told him about the problem. No problem, he said, he had two employees on his payroll who would like to pick up some weekend money and could use a truck. The bottom line is that the move worked out very well, and all because of my dealer network. My father-in-law thought I had done a good job from 350 mi away.

The second example is this. I received a call from a network dealer who lived a couple of states away. He planned to visit my area on his summer vacation. So I provided him with all the information he needed.

This dealer network sideline can be used both for business and personal things, and all because you took on metal building contracting.

Now you are the metal building contractor, so let's go get some work. The first thing everyone thinks about is the bid market, in which you put in a lot of work and time to come up with the cheapest way to do the job. After you turn in your price, you hope nothing has been left out that will create a real problem if you do get the job.

Design-Build

Now the bid market is a constant source of work, and you can participate when you need or want to. So now let's look at your other source of jobs, the *design-build* field of construction.

The design-build concept can be used in any kind of construction, but it seems to really work out with metal buildings in the commercial and industrial business community. Let's examine what we are talking about in detail.

This concept has really taken off in selling metal buildings. I have seen information that puts design-build way over the bid process in metal building marketing.

Concerning design-build construction, you, the contractor, are responsible for putting the entire job together, from coming up with the prospect who needs a building, to making plans, pulling permits, building the facility, and ending up with a happy owner. In other words, you are selling your special services to the business community just as any other professional does.

Design-build requires you to team up with architects and engineers to put together the plans; then you need material suppliers and a subcontractor network to check out special situations the owners come up with before the contract price is established. Then after the contract is signed, the project is given special attention so that you will continue to look good to the owners who, when they end up happy with their new building, tell everybody how good a contractor you are.

You are in complete control of the project, and the owners have only one person who takes responsibility, which keeps things basically simple and easy to work with. The design-build details are covered in Chap. 7.

Once you have your design-build team put together, then it is time to go out and get some business. This is where the dealership with your metal building supplier starts to pay off. You can tie in and use the supplier's ways of advertising in the business community to generate business; and many times the metal building supplier is contacted by the prospect directly and then is passed on to you. I have built some very good jobs that were passed on to me by my building supplier, and if I were not part of the team, I would not have been in the picture.

So you are now a metal building contractor all tied into a dealership, and you are going to use the design-build method to market buildings and make a profit. Notice I said *profit,* not make *money,* because profit is the bottom line, and that's what all this is about. So let's move on and locate the people who need your services.

NOTES

NOTES

2 Locating the Prospective Owner

Nothing happens until you have someone to offer your services to, so the first thing to do is to locate the prospective owner. This chapter will solve that problem for you. I'm sure everyone has heard the old adage, "Gold is where you find it." The same idea applies to construction prospects. If there is one certain phase of construction sales where plain old hard work pays off every time, it's "digging up" the prospects. It's simply percentages—talking to enough people, looking deeply enough in the right places, and keeping your eyes and ears open will always pay off.

The terms *lead* and *prospect* will be used *a great deal,* so definitions explaining what they mean and how they apply to X, the owner, is our starting point.

Owner X is a *lead* before he or she becomes a prospect.

You have watched enough TV police programs to be familiar with the term *lead* and with how leads are used. The good construction salesperson will use leads just as the good police officer does. A lead can be a name or something you saw at someone's business address or a phone call from a friend; leads can come from anywhere. They are the one thing you cannot do without—no leads, no prospects, no contracts, no money. It's that simple. Until you qualify the lead and confirm that he or she is in the market for a new building or some kind of construction work, the lead is not a prospect.

The main thrust of this chapter is to show you how to locate the leads that can be turned into prospects. So please don't confuse the terms *lead* and *prospect.* To reiterate: A *lead* is someone you think *might* need a new facility. A *prospect* is someone you *know* is in the market for a new building.

It's important for the construction sales representative to have a prospect list and to constantly update it. To have a productive prospect list, the salesperson has to check out and follow up on a large number of leads. Many leads result in a small prospect list that produces contracts.

Obtaining and checking out leads is one of the most important functions of construction sales.

This chapter discusses how to go about obtaining those all-important leads, so let's get right into it.

Personal Contacts

This subject can cover a great variety of situations. You're at a party, and a friend comes up and tells you about someone who's planning to build. Or it can be your banker or a sales representative calling on you who passes the word.

Some of the personal-contact leads are just plain luck. That friend you saw at the party might not have taken the trouble to call you at your office. There's nothing you can do about the leads you miss because somebody fails to tell you, but there is one positive action you can take—make sure that everyone you associate with knows what you do (Fig. 3). If, as a contractor, you're going into the metal building field, then let your friends know, both social friends and those in the business community.

Figure 3. Let everybody know what you do.

There is a courtesy I always observe when a lead has been passed on to me personally. After I've checked out the lead, I contact the person who gave it to me and provide a status report. The report may be, sorry, owner X isn't going to build, or she's thinking about it, but her lease has one more year to go. The important part is that you show the person that you took her or him seriously and that you appreciate the lead. At the same time, your friend has the straight scoop about someone else in business, which might make the friend look good by being knowledgeable. This sometimes can be most important, since everybody likes to have inside information.

A fine source of personal leads can be clubs, either social or civic. If you don't belong to any, then I suggest you look into some. It's very important to know the right kind of people in the business community—the bankers, real estate brokers, lawyers, and loan company officials. They are the people located at or near the money, and everything in business is either going to or coming from the source of money.

Remember, it's strictly percentages; the more people you know and who know what you do, the more leads you will have. The old saying "It's not what you know but who you know" is very accurate. For example, a few years ago I closed a contract that came from a neighbor. He told me that his employer had to move, his building had been purchased by the government, and I should talk with him. So get the word around town; it'll pay off for you.

There are two particular groups in which I make it a point to have personal contacts; the first is commercial real estate brokers. Nothing will happen until owner X has a site, so the broker can be a good source of leads (Fig. 4). At the same time, the broker may need your help because owner X won't buy the land before having some idea of what the entire project will cost, and you'll be able to provide those numbers right up front. Develop a rapport in the commercial real estate business, and let the brokers know you stand ready to help them close a deal.

It's really a two-way street, and the contractor who's actively out hunting for business will turn over leads who are thinking about building but don't have a site. Thus, you are able to pass a lead on to a commercial broker. Now you're helping yourself, since when owner X obtains a site, he or she becomes a prime prospect.

There is one drawback when you are passing the lead, however: security (to be covered in detail later). Unless the broker can maintain a confidence, don't use him or her. It's extremely important that the lead not get out on the street. If you dug up the lead, there's a good chance owner X may not be working with anyone else; and the fewer people who know about owner X, the better your chances are for a contract. So work with a real estate broker who will work with you in every sense of the word as well as make money for both of you.

Figure 4. This job was from a commercial real estate connection.

The second group that is a good source of leads is the subcontractors you work with, mainly the heating, air conditioning, electrical, and plumbing companies, since they will be doing different types of commercial work for the business community that is not new construction. A great deal of their work can be repair, maintenance, and renovation. These people will be going into a company and are in a position to notice and hear what plans the company may have. It can be a simple remark. The heating subcontractor asks why the new ductwork is not being extended to the rear section of the building, and the answer is that the section is to be replaced in the near future. That's all it takes, and the subcontractor should be on the phone to you, passing on what is obviously an excellent lead. This also is a good source for metal building maintenance and repair work.

If you have the proper relationship with subcontractors, they really are passing the lead to you mainly to get themselves a sure job—they know that you're pretty good at negotiating contracts, and that if you get the job, they have the subcontractor work in their trade. In other words, they are looking out for number one, but in the process they let you do the same.

Well, there's more to selling than looking the prospect in the eye and stating that you want to do the work. You have to be aggressive in other areas, and that means taking the initiative. Don't sit around and wait for those people who can furnish leads from the business community to give you a call. People are generally busy, and it's very easy for someone with the best intentions simply to forget. So give them a call every so often and ask if they have anything that looks promising.

Learn to pick up information every chance you have, whenever you see people who can furnish leads. (I'm speaking of business friends. I feel everyone has his own manner of mixing business with pleasure. My personal feeling is that it's okay if the other party approaches you, and then you don't stay on the subject very long. If it's something that should be discussed in detail, then make an appointment for business hours, just as any other professional would.) As I stated, learn to ferret out information from these business sources of leads. Ask questions.

Let me illustrate. A real estate broker called one morning and asked if I could give him a budget price. He was trying to put together a lease deal with a hardware company that wanted to expand. The cost of the construction was needed in order to work out the lease amount. Naturally I agreed and turned the price over 3 days later. Two weeks went by; then I ran into the broker and asked how the negotiations were progressing. He told me he'd lost out to another group that also had been making a proposal. There was a time when I'd have told my broker friend I was very sorry to hear he had lost out and walked off. But I've since learned better. When he told me another group had the lease, I had a hot lead to be pursued immediately. I casually asked who the other group was. The answer was some lawyer, and he thought his name was this or maybe that.

I thanked him and went straight to my office and called my attorney. He gave me two names of people who he thought it might be, and on the first call I hit pay dirt. I went to work and presented my proposal several days later. The story doesn't have a happy ending, though; I didn't get the job. Nevertheless, the pertinent point is the manner in which the lead was "dug up."

Remember to dig (I know I've overworked that word, but it's the best word to convey the meaning I wish to put across); practice pulling out every scrap of information when there's a possible lead involved. People know more than they realize.

Referrals

Referrals are different from personal contacts, in that they come from people you've done work for in the past. A referral can also come from another contractor who can, but does not want to, work with the prospect for some reason. This is an extremely valuable source of business because you know the lead is interested in doing something if she or he goes to the trouble of talking to an owner and then contacting you. The problem with referrals is the time lag involved before they begin to pay dividends. You might wonder how this can be, since you have been contracting for years and have plenty of owners out there. But think: What kind of owner? Is it one who says, when asked by owner X about his or her contractor, "Good outfit, but you're going to need your plans first before they'll talk to you"? Or will the owner say, "Good outfit. They can take your project right from the start and do the whole thing for you, a complete turn-key job"? In my opinion the only way to get the second recommendation is to be in that type of design-build work (Fig. 5).

Figure 5. These storage buildings came from a referral.

Most bid jobs have rather long-term project owner relations, and when the problems begin to pop up, the contractor retreats back to the plans and specifications and makes little effort to work with the owner. This has to color the owner's thinking about the contractor, so why should he pass out glowing reports about how great that contractor is?

Job Signs

Take job signs, for instance. Where else do you get the opportunity to put up a sign on the side of the road and tell the world what you do? Yet it seems to me that only the large projects get a decent job sign, and usually everyone who is anyone connected with the job has his or her name on it. What about the average job, say, a 5000 ft^2 retail store? What's out front? From my observations it will be a 2-ft by 3-ft sign that shows its age. A prospect has to get out of the car, jump a ditch, and push aside some weeds to read the phone number. Well, I'm willing to bet that few do.

Granted, many general contractors are fully aware of company exposure achieved through job signs and put up nice large ones. That's great, but how well are they taken care of? They look fine at first, then the weather ages them; or maybe one of our Monday morning truck drivers backs into the sign, and for the rest of the job it's not even straight. Job signs are the best free advertisement you have, so it's important to use a large, well-done sign and then to keep it looking good throughout the entire job. And don't take it down until the very last minute. I've had owners ask me in a joking but partly serious way when I was going to remove my sign! So use them: big, well-done, and looking good.

Along with job signs go construction trailers, and you may use the side for your job sign. This is fine as long as the sign looks up-to-date. However, I prefer the job sign because it stays longer. You can put it up long before you move the trailer onto the site and leave it after the trailer is gone.

In Chap. 9, I present details of working with the owner during construction, but here is the very first step, which I'm bringing up now because it goes hand-in-glove with job signs. When you price the job, add a few dollars to have an owner sign painted and mounted right alongside yours. The first weekend after signs are up, the owner will probably spend all day Sunday riding by to see

New Home of Owner X's Tire Service

You've put the owner right out in front. Owner X is moving up and will love it.

While we're on this subject, think about the location. Start out with the sign in the most advantageous place, even if it has to be moved later. A beautiful sign is worthless if not seen.

Yellow Pages

The next pertinent subject is use of the Yellow Pages. Don't be listed only under general contractors; have a separate block that shows you are a metal building contractor and the services that you can offer the prospect. The Yellow Pages is one of your more successful advertising methods. I find myself turning to them all the time. It's possible to keep tabs on the results from Yellow Pages by the incoming calls, but you're going to have to instruct the person answering the phone to ask the caller nicely if the call originated with the phone book. Study the ads that your competition has, and work up something for yourself. I imagine your local phone company can be of some help.

Incoming phone inquiries are very good leads because the people are coming to you, which expresses interest on their part. At the same time, you or whoever talks to the incoming caller will be able to secure necessary information, and this will save the sales representative time.

If your office is like most, you have someone who handles the phone, and often there'll be no one to talk details with the caller except for the secretary. In this case, train him or her to ask the proper questions so that the salesperson can do some homework before returning the call. Many times the lead is no lead at all, and a smart secretary can again save time.

It's important to find out exactly why the caller is calling. If the caller turns out to be a bona fide lead, then your office help should try to determine the building size and just how the building will be used. The salesman can be in step when she or he calls back, and that's important in helping the lead start to develop a good impression.

A word here about calling back. Do it! Promptly. That's the first good impression the caller will form. I'm absolutely amazed at the number of businesspeople who don't take returning phone calls seriously.

Newspaper Advertising

Have you ever thought about advertising in the newspaper? You may want to look into it. Do some homework with your local newspapers, and get costs, sizes of ads, and the population you will be covering. Although a newspaper ad will reach a lot of people, only a small percentage will be thinking about building a metal building. But it only takes one to make the ad worthwhile.

There is one particular type of newspaper ad that is really more public relations than advertising, but it does let people know what you do. This is the ad that congratulates owner X on opening the new facility. This ad may be one-quarter of a page or even larger. Usually the general contractor can get the subcontractors and suppliers to chip in, and in return their company names also appear in the ad. Maybe some of you have already been involved with this type of advertising. The owner is pleased as punch to see the splash in the paper, and it is good business because a satisfied owner can be your best sales asset at certain times. Admittedly, it's difficult to determine just how successful this kind of ad is. It's a lot like trying to assess the worth of job signs. You know they help, but you're not sure how much.

So take a look at your advertising. Bring in some professional help if you feel you need it. Just remember, when you have something to sell, spread the word.

Direct Mail

This is another area of advertising that can be used. Like newspaper ads, it covers a lot of ground, but the advantage of direct mail is that it will go to people who may be in the market for a new facility, be it now or in the future. Also note that the recipient responding to direct mail shows interest, and this is important. Granted, some of this interest is simply to get information for some reason; but still, it takes only a few bona fide leads to make the mailing worth the effort.

Teaming up with suppliers that promote direct mail is a good method to follow. As a metal building contractor, you may have already been involved with working with your building supplier and may feel comfortable in this area. If so, great; if not, check it out. The metal building industry is very keen on direct-mail campaigns, which is another reason to consider a metal building line as part of your business.

Calling Card

I consider the calling card a very special form of advertising, and for it to be effective, it must inform the reader of exactly what you do. I have seen many cards that listed the company name, the person's name and title, fax and phone numbers, and address—but not one word describing what the firm does. I guess if the word *construction* is part of the company name, then that's all the job description that is needed. However, I feel you should take full advantage of every opportunity to describe what you do; so under the company name, outline any particular area of construction in which you specialize. For example, my card stresses commercial construction and metal buildings under the company name. I carry this two steps further. On my company letterheads and envelopes, the same areas of specialization are listed.

My next recommendation concerning calling cards is, Always have one handy, no matter what you are doing. It makes no difference if the people you are with are a social or business contact. Whether it is a couple of people you are having lunch with or a huge wedding reception, always have a card handy, because if the conversation gets into what you do, it's a simple matter to pass over a card—or better yet, to exchange cards, especially if the card you get is a good lead.

I find the card very handy. Whenever I stop in a business place that has a form to fill out, I put the card on the counter, which makes it really easy for the clerk. While I do this mainly for convenience, you never know when something positive may be uncovered. I have a good example that recently happened. For recreation, my family and I enjoy boating on the Chesapeake Bay. The first thing you do when you visit a marina is to check in at the office and take care of the paperwork. Again to help keep things simple, I use a calling card with the boat's name written across the bottom. Most of the time the card is not returned; and in one case, after I had checked in and was turning to leave, the owner came into the front office, noticed the card, picked it up out of curiosity, I am sure, glanced at it, then asked if it was mine. I answered yes.

He then invited me into his private office and explained he had some serious leaking problems with two of his metal buildings, so since my card stressed metal buildings, would I please have a look at the problems. I did and made recommendations to solve the problems. He then asked me if I could do the work. It was tempting, but a little too far from my working area, and frankly I had all I could do at the time, so I declined. But I did tell him I had some contacts and someone would be calling him.

When I got back home, I called my metal building salesman who serviced the area of the marina as well as my area, and I passed the lead on to him to give to his local dealer for follow-up. So when it comes to calling cards, always have them handy.

Trends in Specific Business Areas

This subject will take a little more work and observation on the salesperson's part. Business trends of all types are taking place constantly. Some are quite obvious, and others won't be noticed unless you take the trouble to look. For example, from 1990 to 1993, when most of the nation was in a recession, the contractor and many other people found it hard going just to keep the doors open. Kindred souls were the automobile dealers. Can you remember when they would do anything to sell a car? While the car dealers were having rough times, as were many allied businesses in the automobile industry, auto parts stores were doing great. People were driving what they had for longer periods; therefore, they needed parts. Also prospering were repair garages, which made money keeping the old car on the road, at the same time buying parts from the parts suppliers.

Suppose motorcycle sales have gone up astronomically over the past few years. There are many reasons why, which are not our concern, but the fact that sales are up makes motorcycle dealers a prime target for leads. At the same time as you are doing your homework on a business trend, make it a point to learn what you can of a general nature about the business. Sometimes you can get help from an unknown source. We're talking about the motorcycle dealer, so consider this: I understand that most started out as second lines for car dealers, and motorcycle manufacturers don't want their cycles marketed alongside cars—they want a completely separate facility for the bikes. Some dealers could be required to separate the two; in other cases it could be the car manufacturer asking for the split. Whichever is true, this kind of knowledge gives you a good source of leads that you probably wouldn't have pursued, had you not taken the trouble to learn something of the industry in which you are interested.

The trend can be going on in any business area; all the salesperson has to do is to dig around. Talk with people who deal with the whole spectrum of business: bankers, stockbrokers, real estate brokers, lawyers, and mortgage loan officers. Talk with friends who are in other business fields. A good place to maintain good relations is city hall. Maybe there is some new regulation that will cause some special construction to take place. For example, you might discover that the health department has a new ruling about the storage of certain kinds of food, and food distributors will become a prime area to be worked.

Trend leads do have the drawback of being somewhat long-range if the salesperson is on the ball. That is, the representative may call on owner X before the owner is really ready to expand. X knows the business is good; and if you caught owner X early, it'll take some time before he or she is ready to build or add on. Once in a while, you'll catch a fast-moving trend lead, but my experience has been that they have to be nursed along for a period of time. How long? I can't say. We're dealing with people; therefore, there's no yardstick. I'll give it one quick guess based on past experience, though: 6 to 12 months.

I consider publications to be a good source of trend information—the better news magazines, financial papers, and trade journals for other fields. I make it a habit to look through any trade magazines that are in the lead's waiting room, and you don't have to be in an office. Just about every business will have trade publications lying around somewhere. Look at them; never forget that leads are where you find them.

Environmental Construction Trends

To carry the business trends concept a step further, let's address environmental construction. This area is beginning to really affect the construction industry and also fits in very nicely with the design-build concept.

Although business trends are market-driven, environmental construction is not. This trend is driven by regulations, be they local, state, or federal, that the business community and the private sector must take into consideration.

Here is where the environmental professionals part of your design-build team can be a big asset. Stay in touch and work with them to see what particular areas can be pursued for business prospects.

When you start making environmental sales calls, you'll probably note the same thing I did: Often the person you are speaking with is not up to date on the regulations and is learning from you that there may be a big problem to take care of. The important thing is there is somebody now in the picture who can solve the problem.

Redevelopment Projects

Redevelopment projects can be a veritable gold mine of leads. It seems that almost every city has a program to buy up old sections of town, bulldoze the buildings, and rebuild. The businesspeople in these areas are being forced to move, making them prime leads to call on and work. I have personally found that many of the older, established distributor types of business are located in these older areas of town, giving the construction salesperson not only an excellent lead but also one that appeals from the standpoint of using a metal building.

The place to start is at the local redevelopment project office. Gather all the pertinent information possible, talk with officials, look at maps, and come away with anything they'll give you. The redevelopment officials will certainly not be able to recommend you to a prospect, but they should have no qualms about telling you about owner X down on Sixth Street who's dragging his feet about leaving. The officials may be good enough to pass on lists of names whom you can contact, which allows you to call first. One very important fact should be uppermost in your mind, though; the lead is faced with having to move. There may be a time element involved, but at some point the lead is going. Rest assured that you cannot find a better lead to work.

Now if you are aware of redevelopment projects in your area and are familiar with them, you may feel there's no need to waste your time talking to the officials. You'll walk right in and meet owner X. Don't do it; always check in with the office and learn all you can. After all, the area you can see being redeveloped may be a small part of a much larger whole that only shows up on some map in the local office. While you are at it, make it a point to stop by every so often and keep up to date on the changes.

Relocation Projects

Relocation projects are similar to redevelopment projects but have one main difference: When you see the work taking place, chances are you're too late to sell a job. The relocating businessperson who has to move will have all the details sewn up before he or she moves out and the bulldozers move in. Again, these leads are prime ones because owner X is being forced to move. No question about it; X will build a building.

Relocation projects bring to mind rights-of-way for new highways or roads. This area offers very productive leads. At the same time, expand your thinking into other fields that might offer leads. How about business areas around a fast-expanding unit, say, a college or a hospital complex? Military installations offer good leads, and I've just had this brought home to me in a very positive manner. The local Air Force base enlarged the safety zone at the end of one of the main runways, and the government bought out three businesspeople. Two rebuilt, and I negotiated a contract with one for a 2600 ft^2 metal building. Point of interest: The other owner had some construction contacts, acted as his own general contractor, and rebuilt; so I didn't lose him to my competition.

You have to work at finding relocation leads. Watch the paper; my experience has been that relocation projects are advertised in the papers as well as carried as a news item. Another way to discover what's going on is to find out the government agencies (federal, state, and local) that are concerned with relocation and then go see them from time to time.

Of course, you'll learn, when you start looking for leads, that one lead-gathering method will help with another; the lead will be the result of the combining of two or more methods. For instance, a friend at a party happens to mention a relocation project—no names, just a lead for a lead, which provides a prime area to investigate.

I repeat: Leads from relocation projects are extremely productive because owner X has to move.

Industrial Commissions

Most larger cities are very interested in bringing new industry to their locale; therefore, there should be an industrial commission that the construction sales representative can contact for leads. Many times the commission operates an industrial park, so that ready sites can be provided to the prospect shopping around for a new plant location.

It will pay the salesperson to check out this area and become acquainted with the officials. I do want to point out one important fact: This type of lead is not as productive as the ones we covered earlier, because the lead in the beginning is just looking around. She or he has no site and maybe is not even sure of the desired location.

Work with these leads only after you know that a site has been obtained. The out-of-town lookers are notorious for taking your time and information and then moving on. Don't ignore this source of leads, but keep both eyes open when you're working with this type.

Rejected Bids

I particularly enjoy working with this segment. I've had several good money-making jobs from the rejected-bid ranks. This source of leads is also an easy area to pursue. Just sit at your desk and look through the material from your bidder's exchange, and then pick up the phone. Rejected bids are not leads but are prime prospects, because the owner has everything together and is ready to go. Then, wham! The owner is knocked over by the prices and wonders what to do next.

Suddenly the construction sales representative appears on the scene and starts showing owner X how to lower the cost and hopefully build the facility without losing needed requirements. A personal experience illustrates exactly what I'm talking about. It happened several years ago while I was working as a construction salesman. A local equipment company had plans prepared for a new metal building facility; it went out to bid and was floored by prices $50,000 to $60,000 over budget. I came across this fact through one of its outside salespeople before it became public knowledge, and I immediately contacted the owner. I picked up a set of plans, and my company studied them for a week with the aim of cutting construction costs without decreasing square footage. At the end of the week, I met with the owner and told him not only could we lower costs but also there was a good chance the final price could be under the original budget. The owner, who was one sharp individual, answered my presentation with two words: Show me! I explained that if he could spare the time, I wanted to arrange a conference at my office so that the estimator and owner of my company could explain in detail plus have the figures handy if the owner wanted to think about other changes.

My reasons for setting up the meeting were twofold: First, I did want our estimator, who was a licensed structural engineer, to explain in detail what was proposed; second, it's a good selling technique to show the prospect she or he is dealing with a first-rate outfit—and what better way than to let the prospect see a busy, well-run office? We showed the owner that the structure he wanted to build was basically a metal building but not of the standard preengineered variety; also, the exterior panels were of a very expensive material. Could he afford that much money for appearance? No was the answer.

Our recommendation was to substitute a standard preengineered metal building with the manufacturer's standard exterior panels. This represented a substantial savings. Another big savings came from the parts department. The original plans had this area air-conditioned because the owner had so instructed. No one had bothered to explain to him that this was a large volume of space to cool and that it would be costly. Instead, we recommended using a mechanical air-change system that would bring in outside air while exhausting the warmer air from inside the area. After all, the shop wasn't air-conditioned, and the parts department was attached to the shop area. Why hand a nice cold part to a mechanic that you are paying to cool? There did have to be some provision for the parts people to be able to work in warm weather, hence the mechanical air-change system. All the mechanics had to do was to open the huge door at each end of the shop for extra ventilation.

Another place where we saved money was in the glass in the front office section. I recall that a special bronze glass was specified, but the front of the building had an overhanging mansard. Also, many large oak trees shaded the entire front. We suggested regular clear commercial glass, since the sun's heat wasn't a real problem.

We suggested quite a few more money-saving ideas, but they were all small, so I won't get into them now. Needless to say, the owner contracted with my company to build his new facility, incorporating our money-saving ideas. Our final price was indeed under his original budget figure, and he didn't lose 1 ft^2 of area. It was also a profitable project for both the company and me personally. The ironic thing was that the lead didn't come from the bidder's exchange but from a personal contact, which proves that business is where you find it, or in some cases, where it finds you.

The point to remember about rejected-bid prospects is speed. You have to move fast. As I said earlier, most negotiated jobs take time, but the rejected bid is the exception. Our owner X is ready to go; everything has been worked out—site, plans, and money in most cases; all the owner needs is some help in getting the price back down to earth. Unless there's a special situation with a lead or prospect, I personally put the rejected bid right up front, with precedence over anything else I happen to be working on.

While on the subject of rejected bids, I made the assumption that the construction salesperson or someone else in the company has the ability and know-how to actually redesign a commercial project, with savings recognized from different construction techniques and types of materials. You will not have much of a chance to sell the rejected-bid prospect if you can't offer some tangible evidence of how you're going to achieve what you promise. If you need help in this area, get it; don't try to bluff your way with the prospect. At some point it will show that you don't know what you are doing, and owner X will give you your walking papers.

Not all rejected-bid negotiated contracts will go as smoothly as the example I just described. Sometimes the owner won't be able to get all her or his requirements. It may not be feasible to bring the cost down to what the owner is looking for without some compromise on the owner's part. When this happens, I push the idea of "better something than nothing." I stress compromise now with expansion designed into the plans to pick up the dropped requirements in the future. When informed of the cost of items in the original job plans, the owner may find there are some that he or she can do without. For example, when owners start moving up in the world, they tell themselves they are going to do it right. So they instruct the architect that they want a large office with adjoining bath. Then, almost as an afterthought, owners add a shower to the bath. Owners are busy people, and many times they would find it convenient to change clothes before attending a club meeting in the evening. The architect says, "Sure thing," and draws in a 20-ft by 20-ft office with a ceramic-tile full bath.

If this job reaches the rejected list, the construction sales representative may have to ask the owners if that big office and bath are really necessary because they're costing so much money. Could the owners get along just as well with a 12-ft by 12-ft office with a half-bath that doesn't have ceramic tile? My experience has been that owners quickly cut out what are obviously luxury features and start working with the hard monetary facts so that they can build the project they want so badly.

The luxury compromise doesn't really cause the owners very much concern; they are businesspeople, and facts are facts when they are faced with the decision of whether the money spent will help them make money. On the other hand, a compromise that may interfere with the owners making money is something not to be taken lightly; it requires some thought.

The future owner may conclude that she or he is going to cut and compromise only so much. The owner's next step will be to accept a higher budget and go after more money. Guess who will be standing right there with dependable, realistic prices plus any other help when the owner returns to her or his bankers? That's right—you!

The one thing the construction sales representative must always remember about rejected bids is to get the job off the bidder's market and then have the owner commit as quickly as possible. Never lose sight of this; the longer the project stays on the street, the better are the chances that something will happen to foul you up. Believe me, it can be some small insignificant point that blows the whole deal.

Remember, the rejected bid is not a lead but a prime prospect, because all the preliminary work has been done. The owner wants to build, so give the prospect top priority and move fast!

Cold Calls

In this particular type of lead gathering, the construction salesperson must be able to think on his or her feet and observe details. Many salespeople don't enjoy making cold calls; they're uncomfortable going in when not expected. To be successful, the construction sales agent should adopt the same attitude that the old prospectors had—"the next rock I turn over could have the mother lode under it." Every door you walk through could have a million-dollar contract waiting.

When a construction salesperson makes that cold call, she or he has been able to observe the physical surroundings. Maybe the agent noticed material stored in the open or obviously incoming boxes left on the sidewalks; or it could be a garage almost covered up with cars to be worked on. In other words, the salesperson is able to observe that the business owner might be in the market for a building, and armed with this knowledge, the agent walks right in and introduces her- or himself.

Cold calls are a necessary source of leads that should not be overlooked. Admittedly, some construction sales representatives regard them as a necessary evil. The physical act of cold calling is often combined with another lead source. For example, the lead may come from a personal contact, and then the cold call is used to see the lead. Granted, this method should be used only as a last resort. If you have the lead's name, it will be possible to call and arrange an appointment. In my experience, when I have combined cold calls with any other source of leads, it has been because I didn't know enough about the lead to feel comfortable calling ahead; but this rarely happens.

Maybe I should give you my definition of a cold call before you skip the rest of this section. A construction cold call is one in which, no matter what the reason, it has been determined that someone may be in the market for metal building construction of some type—new, add-on, or repairs. Please be advised that I'm not suggesting you start at one end of the street and proceed to the other end, knocking on every door as you go. There's just not enough time; besides, that is a very unproductive use of your time.

I've worked out a very special use for the cold call, which you might try: I use cold calls as time fillers during the day. I never set aside a certain time to make cold calls, except in a very special situation; as you well know, in the construction business too much happens on short notice that requires your attention.

For example, let's say I have a 3 P.M. appointment and I've just left a job I'd been checking at 2 P.M. That gives me an hour, which may be lost time, so I'll put that hour to good use. I'll visit a commercial or industrial part of town that I haven't seen for a while and just look around. I simply cruise about, observing the businesses, looking for the telltale signs of growing pains. When I spot a likely looking lead and there's enough time, I go right in and start asking questions. Sometimes, because of time constraints, I have to return to make the call, but the lead is investigated.

I try to make approximately five cold calls a week; some weeks I make more, and others I make fewer or none. The construction salesperson should always keep the cold call in the back of his or her mind and not only plan to make it, but do it! Cold calls require discipline. It is easy to put them off. I know; I do it sometimes. It takes only a weak excuse not to make the cold calls that you know, beyond a shadow of a doubt, you should be making.

A construction cold call is a qualified call; you don't stumble around, trusting blind luck. The successful salesperson knows intimately the signs of a business that is growing. The sales agent is always looking about while driving, rather like a hunter after game. Common sense tells you what to look for, such as old buildings; materials stacked outdoors that really should be under cover (industrial firms display this telltale sign); jammed loading docks and truck parking areas; in the auto industry, many more cars surrounding a building than seems normal (this is a good indicator for auto garages and allied specialty shops such as radiator or body shops); trucks loading and unloading on the street through a front door with goods stacked on the sidewalk.

Another special area to note is metal buildings that are for sale. When you see the building has been sold, then locate the new owner and offer your services.

The details will differ for different types of businesses, and there may be special details to look for in certain areas, say, in the rural business community as compared with the metropolitan segment. But one point will not change: There will be some sort of telltale sign of growing pains, and that's your tip-off to make the cold call. A little practice will go a long way here; and the beauty of it is that you can do it while going from place to place, which makes maximum use of your time. Time is the one thing you'll never have enough of when you really get into selling metal buildings.

You may wonder why you should walk in on owner X. Why not jot down the name and address of the business and call? There is one good reason: the fast brush-off—and not because owner X is rude and hard to get along with. The brush-off I'm speaking of is the kind every salesperson gets while out there doing the job. Our owner X is busy; she's up to her knees in alligators when her secretary tells her there's a Mr. Somebody on the phone about a new building. Remember, the lead or prospect will always put customers first and you second; so owner X gets on the line and tells you she doesn't have time to talk now, but maybe later. Keep calling and catching her busy, and suddenly you're a pest.

Well, you ask, what's the difference between this call and one you are following up from a personal contact? It's simple: Most of the time with a personal-contact call, you have an entry—the mutual friend. When you get owner X on the phone and mention you're calling because Jack Smith suggested it, then owner X will be more inclined to give you a minute or two. You aren't a complete unknown; there is a line of contact between X and the strange voice on the phone. Take that line of contact out of the conversation, and you're next to nothing; the only way to overcome this problem is to let X have a look at you. At the same time, you can look X over and try to learn all you possibly can. If X is busy, you can wait until she has a minute; you'll be able to alter your game plan to suit the situation and make yourself more acceptable. Most businesspeople will take the time to speak to someone who takes the trouble to call on them. So you're able to get to X and start qualifying her to see if she can be moved into the prospect column.

In the event that X doesn't have the time to see you, at least you have a slight entry if you prefer to call before you try to talk to X again. I know it's not much of an entry, but it's better than nothing.

Of course, the lead can be checked out sometimes without talking to the person in charge. If for some reason you aren't able to see owner X, talk with the employees—the second in command or even the secretary. Just a simple statement, such as "X has just signed a new 5-year lease," or "Owner X has taken on a new line and is really going to need some more space," will let you know how to spend your time on this lead.

The best way for a construction salesperson to save time is to find out immediately if the lead is going to build. Knowing when and with whom not to waste time is the surest way to save time.

The cold call is a source of leads that has to be worked physically by the sales agent. You should be aware of one fact about cold calls. They usually are long-lead-time projects simply because the owner probably has done very little to nothing about the project being considered. The exception is repair work; this usually moves quickly.

Also, the cold call can be put in the same category as direct mail. It takes a lot of contacts to generate a small list of leads. This is because you know nothing about it except what you can personally see; it's not like an incoming call from a referral, in which the simple fact that owner X has called you expresses interest on X's part. The redevelopment and relocation project leads both have owners who have to move. You even have some degree of interest on the part of the lead with personal contacts and referrals. But with the cold call, you have nothing until you go in and find out. Simple mathematics tells us it takes a large number of cold-call leads to generate prospects that turn into contracts. The cold-call contract takes more work as a whole, but it's still an excellent source of leads and should be used.

Now let's talk about a very special area of cold calls, which I'll refer to as *prequalified*. These are leads that offer a better chance of results because of a particular situation. I have a good example. A few years ago I realized there are a lot of metal buildings starting to show their age. Therefore, the retrofit market was one that needed to be pursued in an aggressive manner.

In the course of moving about during the day, I took note of older metal buildings and then grouped them into areas. It's easy to spot the signs of age, that is, rust, ding marks, faded paint. It's amazing when you get into something like this how much you know and remember about your area of operation concerning buildings you did not build, which is a big help.

So after a few weeks I had a list of older metal buildings grouped in areas of convenience. From past experience, roof problems were going to be the main complaint on the older metal buildings. Knowing this, I waited until my location (which is coastal Virginia) had a good northeaster off the ocean with howling winds and heavy rains. The very next day, as the sky cleared, I devoted a whole day to calling on the buildings on my list. I was after reroofing jobs. The results were two repair jobs, a good number of future leads to stay in contact with (some have since called me for repair work), and a reroofing job that led to other work. All in all, it was a good day's work of cold calls.

Now back to my example. This firm was a high-technology machine shop doing work for the government, and leaks were more than a mere inconvenience. When I walked into the lobby, I asked the receptionist to let me see someone in management about their roof leaks (assuming they had some). Their problems were so bad that I was talking to the top man in short order. In the course of the conversation, I was asked how I happened to stop in, because the top man assumed at first that someone in plant engineering had called me. I told him the call was a "qualified cold call" looking for business. Frankly he seemed pleased someone thought enough of his business to seek it out. I got a contract to reroof part of the building; then 14 months later I built a 5000 ft^2 model shop, and 11 months after that a 10,000 ft^2 addition to the main shop.

Learn to observe places of business, and then learn to notice changes that occur over a period of time. Always make it a point to keep returning to the business location you feel offers cold-call leads, so that you might see the changes that take place over time. Also, remember to get out there and diligently make cold calls!

The Secondary Cold Call

I know you're wondering just what this is. It's a very special use of the cold call, and the word *secondary* is used because the actual cold call is the result of another action.

I'll start with a direct question. Have you been in a hardware store, or maybe an auto parts store, or say a lumberyard or any type of business along these lines recently? That is, I am speaking of businesses that need fair to good-size storage space? If the answer is yes, did you seek out the owner or manager and ask whether there were any type of expansion plans for some reason?

In other words, you're buying something and when that's done, you make a cold call. Don't laugh. It works, as the following example illustrates. A few years ago, one of the electric eyes on our kitchen stove broke. The next day I stopped in at a local appliance parts dealer and bought a new eye. Then I asked to see the owner or manager. In short order, I was talking with the manager about expansion plans. I told him what I could do for him, and he was quick to point out he needed more storage space badly, and told me to contact the owner whose main office was located at another city 25 mi away. The local operation was one of four branches.

I called the owner and stressed the fact I was calling at the suggestion of his manager (please note I'm not just a voice on the phone, but someone known to one of his managers, which gave me an introduction). The owner explained to me that more storage space was indeed needed, but one thing stood in the way. The problem was that the land behind the store, although owned by him, was not zoned for business, and he had no idea if it could be. I told him that I had some experience in this area and I would be glad to help when he was ready. I said I would stay in contact in the coming months. (Notice I didn't say, Call me when you are ready. I wanted to be the one to check things out.)

Anyway, we worked out the zoning problem, and I built a 2500 ft^2 storage building behind this appliance business, and all because I made a cold call after buying a stove part.

Figure 6 is the appliance storage building I was just going over. The building is positioned and shaped as you see in order to handle two problems. First, the brick building has all its roof water flowing to the back, and the heating and air conditioner units were on the ground behind the building. So the new metal building was pushed back about 5 ft with an entry hall installed for access; note Fig. 7.

Second, the new building would be exposed to a lot of leaves, and gutters would be a maintenance problem. So a single slope was used to get all the roof water to the back so that one-half of it would not have to be handled with the water off the brick building, and no gutter was installed. So the water goes to the back out of the way of everything, and there is no gutter maintenance to worry about. And everything has worked out very well.

Another secondary cold call is shown in Fig. 8. The building to the left is a florist in the town where I live, and one day a couple of years ago I stopped in to order some flowers. I knew the metal building was about 20 years old and probably had a few maintenance problems. So I asked to speak to the owner and talked about what I could do. The roof was leaking badly, so I went up the next day and inspected it plus took some snapshots. The bottom line was that I was to reroof the building.

Figure 6. Secondary cold call—an appliance storage building.

Figure 7. Secondary cold call—new metal building.

Figure 8. Secondary cold call—special photography company.

Now while I was checking out and inspecting the florist's building, I made a cold call on the building to the right in the picture, which was about the same age as the florist's building. Its roof was in bad shape also. And I was told to come up with a price. The building was used by a special photography company, and leaks could really create a problem.

Since both buildings were side by side, it could be handled as one job, and I could lower the construction cost, if both owners would agree to having the work done at the same time. I presented this idea to the owners and was told to get it going, and I had a nice reroofing job of approximately 6000 ft^2.

So, make the secondary cold call when possible.

Disaster Leads

We've all heard the adage "It's an ill wind that doesn't blow somebody some good." Well, truer words have never been spoken with respect to construction salespeople and disasters involving commercial buildings.

Fire is by far the main disaster that strikes buildings, and then Mother Nature can get upset and unleash a tornado or hurricane. No matter what destroys or damages a structure, get yourself in the picture as soon as you can, considering the owner's personal feelings in what must be a trying time.

Ninety-nine times out of a hundred, it's imperative that the owner start a new building or repairs as soon as possible. Here's a good example, which shows what the aggressive sales agent can accomplish in going after a prime disaster prospect. When I worked for a design-build construction company that employed another salesman as well as me, my colleague was working with a man in the furniture business who wanted to open a new branch. The lead, by the way, was a personal contact, which came from a club meeting.

Anyway, this salesman worked like crazy to sell the job. There was another club member after the job also, and the owner chose the other company's proposal over ours. I remember the reason—the other company's price was a couple of thousand dollars lower, and the owner opted to save money. That's what the owner expected, anyway; but well into the job, it suddenly came to light that there was quite a lot of fill dirt needed, and the cost wasn't included in the contract price. The end result was that the owner paid more than the price my firm quoted, since we had included the fill.

The owner was good enough later to admit he should have paid closer attention to details and signed with my colleague. Six weeks later, on a windy Saturday afternoon, the furniture man's main store burned to the ground. That night my colleague called, offering to help in any way, and this time he did build the new building.

This is also an excellent example of remaining objective and not getting personally upset when you lose a job. You never know when that owner will become a prospect again!

Disaster leads can be good ones because the prospect will have to have a building from which to operate. Follow the news media and pay close attention when you hear or see news that may give you a lead.

This is one area in which speed of construction is very important, and the metal building offers fast construction, so this can be used to sell the metal building idea to the owner when the original building was of standard construction.

It's been my experience with my building manufacturer that the disaster job is pushed ahead of the normal schedule, which really speeds things up, and then the building is erected quickly by putting a large crew on the job. The bottom line is that the metal building contractor can give the disaster owner a new building fast. So when possible, use this advantage to get the job.

Bonus to Employees: Finder's Fees

Make sales representatives out of all your employees. Stress to them the importance of bringing every scrap of information that may be a lead to the attention of the proper person. Instruct your people to keep their ears and eyes open. It's positively amazing what they can bring in when they are on the lookout for leads.

You can give employees the old pep talk about how much the leads mean to the firm and what's good for the firm has to be good for them. When the first lead comes in from your bookkeeper and ends up in a big fat contract with a good margin of profit, you could give her congratulations and a big thank-you in front of the whole office staff. Months later you might still be trying to figure out why the hired hands haven't brought in any new leads since that last great one.

Well, nothing in this world is free. Your employees are paid to do a job, nothing more. Why should they go out of their way to put money in the boss's pocket? Their check arrives regardless of the leads. The solution is to make it worth their while. Give the bookkeeper a big thank-you *and* a check in front of the office staff, and notice how there'll always be a lead or two coming from the help. Nothing else, absolutely nothing, motivates as money does, so use it to make more money.

Don't be selective; let everyone who works for you know that he or she has the same standing offer for a lead that ends in a contract. Here's the catch: The lead has to become a contract. This idea of a bonus to employees will work and will provide the firm with a new source of leads because seldom will the employees associate with the same people that the owner does.

Another paid source of leads can be *finder's fees.* In this case you have an understanding with salespeople in other fields, say, equipment sales agents, or maybe a commercial real estate broker, whom you will pay a set amount for a lead that turns into a contract. These people are in and out of many businesses and know what's going on. For example, why are certain businesspeople talking with the equipment sales representative about a new 10-ton bridge crane? Why are they talking to the real estate agent about buying the adjoining property? It would take a real dummy not to be able to see business expansion here.

Others have knowledge that you lack. Realize that you might get your lead as the result of a personal contact, and that's great. But personal contacts are personal; the person with the lead is doing you a favor by telling you—there is no incentive most of the time except for helping a nice person or a friend. If you could only know the leads that were meant to be passed on to you, but fell through some crack for one reason or another, you would just be sick. For every one you get, I personally feel three or four get dropped. The answer could be a finder's fee. Nothing else motivates as well as money. But you can't go around offering money to your friends for leads. You can, however, offer it to people in business whom you see mainly during business hours.

I personally use the finder's fee offer very sparingly because I don't like the idea of being known for having to buy my leads. This really is a personal judgment situation for the construction sales agent. In other words, it's hard to tell someone how and when, if at all, to use finder's fees. There may be accepted business practices in your area that can help you decide on their use.

My own guideline is that when someone brings me a deal, not a true lead, and it puts me in the know, then I'll build in the fee. For example, a person brings me a possible lead, such as that he knows about a new building, and he's trying to sell some part of the overall picture to owner X, and could he make a few bucks if he gets me in to see X. In other words, the lead seller is shopping before the fact, and that's only good business—for me and for him. When I know what the situation is and have some measure of control, I will consider a finder's fee.

Zoning Changes and Deed Transfers

Watch the paper and/or check with city hall for applications for changes of zoning to accommodate commercial enterprises. These are good leads because someone is planning to build and is going to considerable trouble to have the area rezoned.

Deed transfers are another good lead source. When owner X puts out hard cash for land, you know it's serious. Often deed transfers are recorded in the local papers. I'm sure it's different from locality to locality, so you'll have to find a way to keep updated on who buys what. Maybe your area has a publication available like the one where I work. It's a booklet that the businessperson can subscribe to that lists deed transfers as well as lawsuits, who is filing for bankruptcy, and so on. It's a very handy item to have around.

These two lead sources can be pursued by your office help. Let someone be assigned the job every day to check the local papers and business publications and to mark the leads. The important thing is for the checker to cull out the useless ones, leaving the sales agent the job of boiling down the original list. Every now and then leads will come from news items, and these should be brought up also. You'll go for days with dead ends and then bingo! This source of leads really doesn't take too much of your time.

Records

The world's worst record keepers are sales representatives. Most hate paperwork, so they have a tendency to do as little as possible and operate out of their shirt pocket. I'm no different.

It's quite important to have and keep up-to-date records to ensure that no lead or prospect is overlooked. The very fact that long time periods are an integral part of selling construction makes good records mandatory, so that the prospect can be contacted on a regular basis according to the particular situation. It's equally important for the owner of the construction company who has a salesperson working for her or him to know what is going on. Employees leave, and there must be some record so the job can be carried on.

A word of caution concerning your computer. Do have some security control over your database. Prospect information should be somewhat privileged.

It does no good to keep records if you don't use them. I'm into mine everyday, making notes and making sure nothing is being overlooked. Failure to follow up tells the prospect you're not interested in the business, and when the prospect gets this idea, you won't have it.

Careful Consideration of Committees

This subject addresses committees, which is an area that can give the design-build contractor some very special problems. This area has to be approached and handled specially since you are dealing with a group of people instead of an individual. The person in business has a pretty good understanding of what's going on, and the salesperson can respond easily. But everything changes when you have a group of people, most of whom do not have the foggiest notion of what is going on in the construction world.

People with the best intentions in the world (since they will be nice people trying to do as good a job as possible) will bankrupt you if you let them. I'm talking about church, social, civic, and even theatrical groups. I firmly believe the first time in history that three or more people met together, the first topic on their list was a building to house their enterprise. It still has to be the main subject of committees to this very day.

The first thing a committee that is interested in a building wants is a cost number. So you attend a meeting to determine the requirements in order to work up a budget.

There are some "tricks of the trade" I would like to pass on for when you find yourself sitting in with a committee. First, take notice of who is the leader (and it does not have to be the chairperson), and then make sure this person knows that you are a construction professional; you in essence will be selling to this person, and then this person will convince the group. At the same time, do not ignore the other members, and the best way to do this is to listen to them when they are making a point. Doing a lot of listening, and making notes will make you look good.

One meeting should give you enough information to work up a budget, and I suggest this be a fast budget simply done in the office with information you have readily available. It's now your call to present the budget at another meeting or let the committee look it over without you. Be on guard to watch your time.

Trying to sell design-build construction to a committee is very time-consuming, and frankly the contract potential is not very good. Many times someone on the committee will think the best way to go is the bid route, and all your work has been for nothing.

I recommend you give a committee lead very careful consideration, and look out for number one.

Contract Potential

Up to now, this chapter has covered where to go to find leads. But that lead is nothing more than a name until you talk with owner X and dig out the necessary information to determine just what it is you have to work with.

The good construction salesperson would make just as good an investigative reporter, police detective, or intelligence agent. Why? Because people in these professions have to be able to obtain information vital to performing the jobs. Information is the job.

As a construction sales agent calling on a lead, you have to ask questions. Never hesitate to ask the particular question that you feel will help you. The very worst the lead can do is to say no, and you're no worse off than you were before you asked. You have nothing to lose.

I'm talking about legitimate questions pertaining to the project; don't be guilty of prying. As long as you're on the subject, ask away. An important fact to keep in mind when talking with a lead is that owner X may be able to pass on another lead regardless of whether X is a bona fide lead. Train yourself to bring the conversation around to your asking if X knows of anybody who may be in the market for a new building. You'll be surprised by the answers, and this type of lead from a lead will usually give you one that probably no one knows anything about yet, which puts you in the picture right up front.

Good information-gathering ability is not something you are born with. It has to be learned and cultivated. When X tells you he doesn't own his building, you ask if the lease is short- or long-term. If the answer is short-term, ask how short, and explain that building a new facility is not a fast undertaking. It takes a lot of time. Since he has, say, 8 months to go, why not let you work up some quick budgets so he can start to get an idea of cost? If the answer is long-term, say, 3 years down the road, ask if he has any plans for a branch operation. Don't just accept his answer about leasing and walk out. Dig to learn something that may help you now or in the future.

In the next chapter I go into detail about what to ask, but for now it's important to realize that the construction salesperson must be able to ferret out bits and pieces from the lead. It's the only way to determine if the lead can become a prospect, and prospects are what you are after, because they are the ones who buy.

Let's quickly sum up this chapter, which is strictly about finding leads to investigate. The total number of leads that you'll be working with at a given time will be made up of leads supplied from the various sources I have discussed. No one area will provide you with enough leads that you can afford to ignore the remainder. Temporarily, under certain conditions, you may get more than is normal from one particular class of lead, but don't let this lull you into a sense of false security. The picture can change, and you'll start out one day with no one to call on.

Keep all the sources constantly in mind, and pursue them. Remember that you'll be playing a percentage game, and it's vitally important to remain completely objective about the source of your leads. If there are things you personally don't care to do, then do them anyway. The reason you're doing this at all is for money, so don't be foolish enough to pass up leads that can provide money.

Cold calls seem to be most construction sales agents' Waterloo. For that matter, cold calls in any field of selling seem to be the one least liked. I've even been told by some salespeople that they feel cold calls are somehow demeaning. To me, this is rubbish. What's the difference between a cold call and a clerk in a retail store coming up and asking, "May I help you?" When the construction salesperson makes a cold call, he or she is asking the same question—maybe not in the same words, but the meaning is the same.

Many salespeople don't like cold calls because they are afraid of them—afraid of looking stupid, afraid of a situation that they can't handle, afraid of being rejected, and so on. Cold calling will teach you to think on your feet, to handle situations, to meet complete strangers in all walks of life, and to talk to them as well. In short, it will give the construction salesperson confidence, without which he or she couldn't give a job away. And after you become comfortable with cold calling, don't stop, even when you have plenty of work. It's a sure way to keep yourself fine-tuned to people. By the way, if you pick up a lead, then great; there is nothing like a good lead to perk up your day, especially when you find that your closest competition knows nothing about it. So wade right in, and worry more about making money than about getting your hands dirty.

Lead gathering is plain hard work; please start out with that in mind. Ask questions, and always be on the lookout for a lead from a lead. Never leave owner X when you are through with your initial meeting without asking if she or he knows of anyone else who might need your services. It's truly surprising what people know but are not aware of until someone asks them.

In closing this chapter I've created a chart of lead-gathering areas along with their contract signing potential for the prospect. I know it seems like a lot to remember, but after you have gained some experience, the chart will be imprinted in your head with your own variations added.

Lead source	Contract potential
1. Personal contacts	Good because lead has expressed interest.
2. Referrals	Very good because lead has expressed interest and taken the trouble to investigate.
3. Cold calls	Fair because you know very little about the call at first. Will take many calls to produce good prospect.
4. Trends in specific business areas	Fair to poor because you know very little about the call at first. Will take many calls to produce good prospect.
5. Redevelopment projects	Excellent because the company has to relocate.
6. Relocation projects	Excellent because the company has to relocate.
7. Industrial commissions	Fair because lead can be checking out many locations.
8. Rejected bids	Excellent because lead has everything together and is ready to move, but cannot do so for some reason, usually money. (Note: Move fast, even at the cost of another lead if necessary.)
9. Advertising	Good because lead has expressed interest.
10. Disasters	Excellent because company will probably need to rebuild to continue business.
11. Bonus to employees; finder's fees	Fair because the person supplying the lead most of the time really doesn't know very much about the lead.
12. Zoning changes and deed transfers	Good because the lead is putting out time and money to do something.
13. Committees	Low contract potential. Be careful.

The lead sources are listed in the chart in decreasing number of leads generated; that is, you'll have more personal contacts than referrals and more relocation projects than rejected bids. This rating system is very loose, but I want to give you some kind of handle for judging your leads, so that when you have to sort through and put priorities on them, you'll have some idea of just who's hot and who's lukewarm. In other words, don't break your neck over a lead from the industrial commission when you have one from a redevelopment project and there is only time to pursue one.

Please remember that leads are the lifeblood of marketing construction. They are where the prospect comes from; and the prospect is the person who hands you the check. So be on the lookout at all times for leads.

NOTES

NOTES

3
Qualifying the Prospective Owner

Chapter 2 covered many methods that can be used to obtain leads. The problem with a lead is that it's nothing but a name on a list until you take the time to check it out and qualify it, to see if the lead can become a bona fide prospect.

Not every lead will become a prospect. Again it's a question of percentages; a lot of leads will boil down to a few prospects. I have absolutely no idea what sort of ratio you're supposed to get when comparing the number of leads to the number of prospects. There are far too many variables, and I feel a statistician would be hard-put to arrive at a number. This much I do know: A lot of leads will give you a few good prospects. Please note the word *good*. A lot of leads can give you a fair number of prospects, but we're not interested in just any prospect—we are after the good ones. These good ones are the leads that you prove to yourself are going to make a move; they plan to build. Granted, there may be problems that must be overcome before the contract is signed; nevertheless, good leads plan to have some type of construction done. That's my definition of a good prospect. All others are nothing but people to whom you very nicely contribute your money and time. Be prepared to work for your prospect list. Every now and then, you'll back into a good one without much work, if any; but that's going to be the exception, not the general rule.

When you are qualifying a lead, use the direct approach. Come right out and ask owner X what his or her plans are. Does the owner intend to do something? You will receive one of two answers, no or yes. If the answer is no, then you know right away this particular lead is a dead end; but never leave without asking X if she or he knows of any other likely people whom you might call on. Remember, never pass up a chance to fish for a lead.

The no answer is cut and dried; you know immediately where you stand. When the answer is yes, then you think you have a real, honest-to-goodness prospect to work with. Well, I hate to burst the bubble, but this isn't necessarily so.

Our owner X who's just answered your questions with an emphatic yes could be telling you yes while in truth not having a clue about what's involved. Owner X may have a burning desire to build that new facility, and just because X wants to, X thinks she or he can do it. It's up to the construction sales agent to dig out enough information from the lead to determine whether the lead is a good prospect.

You won't know when the lead answers yes if that yes is good or bad. The best technique is to assume it's a bad yes and start right off digging for the needed details. If the yes is a good one—that's it, owner X has the preliminaries put together, and it's quite obvious to the sales representative that this lead knows what he or she is about—then you lose nothing at all at the beginning by assuming it's a bad yes. You will quickly pass through the preliminary qualifying and get down to the nitty-gritty. On the other hand, most of the leads will require qualifying from square one, so we'll start at the same place.

Site

The first requirement for a contract is that the lead must have a site. Absolutely nothing will happen unless owner X has a piece of land on which to put the building, be it a new building or simply an add-on to an existing building. Of course, the site is not a requirement for maintenance or repair work, but most of the time it will be an important part of the process. This is as important to selling construction as simple basic addition and subtraction are to the science of mathematics. I stress this point so strongly because I've noticed that construction sales agents have a tendency to become so wrapped up with a project that this simple, basic rule is overlooked.

While you're talking with owner X, you steer the conversation around to the site, and X tells you she doesn't have one yet; as a matter of fact, she hasn't really started looking around, but she's going to do this and that when she locates the right spot. When you hear words to this effect, realize that X is not a good prospect and do not devote very much time to her, if any. I'm not saying you should drop this lead; I'm saying that you should back off until the lead is serious, and lining up a site is the first step in this direction. Keep the lead on your list and check back every few months to express interest and keep your name in front of owner X. After all, the lead can suddenly become serious and buy a site, and you'll want to know about it from the owner, not from the grapevine well after the fact.

I'm sure you're aware that nothing is simple anymore, so there are some variations of this no-site rule to be aware of and to know how to use in your selling. What about the lead who has no site but, you feel, is not daydreaming? The lead is going to do something. I want to be very clear on this point. I'm not saying to forget a lead with no site; I'm saying, don't spend any time or money except what you can easily provide with no trouble to yourself.

A lead came to me in the form of a personal contact. The lead was thinking about building a new facility, so I immediately jumped on it. It didn't take me long to determine that owner X was going to build; the problem was that he was undecided about just where to relocate—hence there was no site. I told owner X there was little I could do until he had a site because the prices had to cover site work. He did give me the general specifications for his new store and asked for a handle on the cost, just a ballpark price. I have done quite a few jobs like the one he wanted, so it was only a 15-minute job to budget the cost. I didn't tell owner X this; I left him with the idea that it would take a little time but I was willing to do it if I could work it in during the next couple of weeks. I wanted owner X to think I was really putting in time to help him. I was simply selling myself. Now that's all the pricing I was going to do—I didn't care what else was asked of me.

I returned with the budget a few days later, spent a few minutes going over it, and then left, telling owner X to let me know when he found a site and that I'd be glad to look at it to evaluate hidden construction costs. This would keep me in the picture and would begin to make owner X dependent on me.

Six months went by; then owner X called me to check out a site with him. I might add that during the 6-month period I had called him several times to touch base so that I could stay in the picture. Well, we checked out the site, and it had nothing but problems—sewer, setback requirements, plus a road exposure that was terrible for a retail business. Site costs would run up the overall costs, and maybe he should look around again, I told him. This went on for a year, and all I had invested was a little time. I felt under the circumstances that the small amount of time was worth it. At no time did I do any more estimating. I was positive owner X would build when he found the correct site. Also, he was calling me, which showed interest. Well, after a year he finally located a site that filled the bill, from his business viewpoint and with respect to construction cost. The real estate agent was a friend, so I asked him to keep me informed of progress made on buying the lot.

Three months later the agent called and told me that owner X had just closed on the site. (This brings up a point that's a complete mystery to me. Why does it take so much time to settle something that's on the market to be sold? Land deals seem to involve no concept of time.) With the site in hand, owner X became a prime prospect, although clearly I had somewhat of the inside track by virtue of my prior work. To make a long story short, it turned out to be a nice job.

My point is that owner X didn't have a site but did meet all the other requirements, so I continued to follow up and work with him as long as the costs to me were rock-bottom minimum. I wasn't going to provide plans, price letters to banks, and give all the other help necessary to get a project off the ground.

I mentioned earlier bringing the interview conversation around to the site. Just how the sales agent goes about asking for information is important. Now with this in mind, imagine how you would look to owner X if, when you asked the owner if she or he had a site and were told no, you closed your briefcase and turned to go, saying over your shoulder, "Call me when you have one." Qualify a lead like that, and you can bet the owner will cross you off her or his list.

The correct way to do it is to build the correct question right into the flow of conversation, for example, "In order to work up a meaningful price, I'll need some information on your site." When you cite the reason for needing the facts, please remember owner X will not be familiar with all the little details but will start to realize there's more to building a new facility than just saying, "Let's do it." You are selling the owner on your competency. So you continue, "It's important for me to know where the sewer is located so I can check with the city to see if there will be any problems tapping in."

The reasons given to the owner for asking for site information can be numerous; just use something. Surprisingly enough, I've found that most of the time there really are legitimate questions that need to be answered.

At this point the owner tells you that a building lot has not been located. Don't let on for one second that warning bells are sounding inside your head; calmly go on with the meeting. Move ahead to the other qualifying questions. Make no hasty decision yet concerning this prospect. You need all the facts first, so go ahead with the digging.

After you're sure nothing will happen until the owner locates a site, you have to pull back without giving the impression to owner X that that's what you're doing. The tack I always take in this situation is to tell the lead that until a location is picked out, it's very difficult to work up any costs that will be worthwhile. I say that construction prices are always in a state of change. It will do no good to supply building costs without site costs because the bank is interested only in the total cost. Then I throw in the hooker: I say that I would be doing the owner a disservice to give out what will become worthless and, worse yet, dangerous information.

At this point I offer to give a handle on the building cost if the owner will use it only for personal information and will take note of the fact that it will be on the high side. Now I offer this mainly to keep myself in the running if, in my judgment, this lead has promise. This judgment call is just what the word says; it's strictly based on my feeling about the owner, how he or she comes across, what has been done to date—in short, the total of all the small parts that come out in a qualifying interview. Here the construction salesperson is completely on his or her own. The agent makes a decision and then proceeds from there, often hoping she or he has done the right thing. This is quite hard to get out of a book. I can try to give you some guidelines, but that's all; the rest is up to you. And if you're working at it, you'll make mistakes. Don't let them worry you. Mistakes come from doing something; mistakes aren't made if you're doing nothing.

Sometimes the fact of no site can be used to help sell yourself to owner X. I think many times our lead hasn't done anything about the site because he or she really doesn't know how to go about it. The lead has some ideas, but may hesitate to become involved with a real estate broker. Many times the lead's so wrapped up in business that she or he won't take the time to drive around and see what's available. Here the construction sales representative can be of some help. Tell the owner some of the details concerning water, sewer, gas for heating, and so on, and then recommend a good commercial broker to work with. If owner X agrees to talk with the broker, offer to call the broker yourself; tell the owner that you'll be able to give the broker construction facts that will help the real estate person do the homework, so as not to waste the owner's time with unsuitable locations.

This strategy does several things for the sales agent. It lets the agent stay in the picture and enables her or him to call the broker with whom he or she wants to work. You give the broker all the information concerning owner X, if possible even going so far as to give the broker your readout on how the owner likes to be approached. The really important fact is that you have called and given the broker a good lead. Since we all travel a two-way business street, the broker will want to repay the favor in the future.

One more excellent point to bear in mind about contacting the broker is the possibility of developing an ongoing relationship that will keep you informed about the progress being made in finding the site. The broker might even learn if another contractor comes into the picture and pass that information on to you. Forewarned is forearmed. Be sure to ask the broker to contact owner X right away to inform X that the broker is starting to work on the site. Then several days later, check with the owner to find out if the broker has indeed made contact. There are two reasons for checking back: to keep the ball rolling and to show owner X that you're a hustler, which is just another small step in selling yourself. Never let up on selling yourself; it has to become as natural as breathing.

Without the site there can be no contract. If the lead doesn't have a site, then be careful about spending time and money on the lead. Do offer to help by bringing in a broker (whom you can work with); then step out of the way, but not out of the picture. And keep tabs on the situation, because when the owner obtains a site, it's time to go to work.

Finances

The second most important fact in qualifying a lead is money. I've always heard that love makes the world go around. I don't believe that—it's money!

Without money the lead is a dead end. It will take money as well as a site to make this lead into a prospect who can sign a contract. There is one major difference between the two. The site location can be critical—not any old place will do—whereas any old money will do nicely.

Bringing up the subject of money in the conversation is a little tougher than bringing up the site. The best approach to opening the door on the money question is to assume that the owner has no problem at all. The owner will need some information to satisfy the lending people, so you offer to help with what's needed. This will usually start the talk about finances, and from there it's an easy matter for the sales representative to ask some direct questions under the guise of gathering information to help. You are doing just that, but at the same time your prime concern is whether owner X has all the ducks in a row so he or she can pay the bills.

Nine times out of ten, the lead will have talked with her or his banker and will have some idea of what to expect. Usually a good businessperson will talk to the banker first, before proceeding further. Without plans and contract costs to supply to the lending people, everything is verbal and tentative. Nevertheless, the lead will have some idea about the money. My experience has been that owner X won't hesitate to tell the construction salesperson about already talking to a banker and that there is no problem. I've also found that's just about all owner X will say about the money; the owner has it lined up, and that's all you will find out until she or he knows more about you personally.

The gray area is what you have to be prepared for now. It's sort of like the twilight zone where it's very hard to prove something that's important to you. Owner X tells you that she or he wants a building, and you ask for details, which are forthcoming. The owner tells you there is a site, and you ask where; the answer is a simple one—you are dealing with something tangible. But how do you prove to yourself that X has the money arranged as indicated? No one likes to be doubted with respect to money, so you'll just have to accept the owner's word.

The only defense you have is to try to find some circumstantial evidence that could give you some idea of whether owner X indeed has the money. The best means is to offer to supply some needed information that the lenders will require. Sometimes owner X will start to discuss some of the financial aspects. Hopefully you'll pick up enough to put your mind at ease. In any case, you'll be in a gray area until the project moves to the plans phase. The good construction sales agent learns to live with this problem; it comes with the territory. It certainly won't hurt if you are able to make some discrete inquiries as to how owner X stands financially within the local community.

Owner X has the money, or at least you have been told so. But what about the lead who has a site and no money? You'll come across this situation every so often. I'm talking about a lead who wants and needs a new building, has optioned a site, but for some reason can't get the money to make the deal fly. The lead will probably not hesitate to tell you his or her tale of woe, for the lead is desperately casting about for help with the problem. This type of lead can be a good prime prospect under the right conditions.

The construction salesperson will have to know something not only about design-building but also about leasing the new facility back to owner X.

You may be asking, Why not assume that the lead has the money to build and not use this aspect as a qualifier? Well you can, and a great many times it will be perfectly all right. But you have no defense against the big talker who likes to think of him- or herself as a high roller. I use a handy rule that is accurate almost every time: Beware of the easy sale and the lead who casually brushes aside talking about project finances.

Money is the reason that you're out there talking with this lead, so find out all you can. But at the same time remember this is probably the touchiest subject you'll discuss with a lead.

Reason for Building

Our person has a site and money, so let's look at another important area. Why do owners want to build? They may be forced to by being in the path of redevelopment and relocation projects; or they may be forced to expand because their competition is expanding and they are afraid of losing business. It could be that their lease is up and they can't renew it. They may have no choice but to build the new warehouse to be able to handle the new accounts they've just taken on. Regardless of the amount and type of force being applied to the lead, owners will only move when they have to.

It's important to the construction sales agent to find out what the moving force is. It will help qualify the lead, to see if there is a good prospect to be had, and help the salesperson determine the time factor involved if the lead does look like a good prospect.

It's vitally important to know the details of why the owner has to build. It helps the sales agent judge exactly how to proceed with the project, and at the same time the salesperson can determine just how serious a lead is. As in any other selling business, you get the lookers—the people who may be thinking it would be nice to have a new building and who want to know what it would cost on today's construction market. This leads us into the next topic.

Interest

There is a way to tell if a lead is really serious or out window-shopping. There's not too much to it: You give the lead some small duty to perform. The lead is serious if he or she does it, and not serious or at least questionable if he or she doesn't do what is asked.

The sales agent should be able to get a feel for the lead as far as interest is concerned; this will come with some experience. It helps if you can have a readout before you have to "assign" a duty to the lead. The problem is that this requires a return call or maybe several until you can make your determination.

Okay, what do you do? Simple. Ask the lead to get you some information on the site from the real estate broker, or the surveyor if one is involved. Ask the lead to check on a zoning question, or perhaps a decision about some detail of the building that you need to know before the price can be estimated. Here's a sure method that I've found most reliable: I ask the lead to give me an hour's time so I might show the lead some buildings, to get a better idea of what we're talking about. If this is really impossible, and it will be with a retail businessperson once in a while, give the lead directions and suggest that she or he drive by on a Sunday afternoon and have a look. If the lead never does, then forget it.

Let me give you a good example. I was trying to qualify a lead who thought he wanted to build a new facility. Everything checked out: He had a site, and the money was there. But owner X didn't seem to want this new building badly enough; something didn't ring true, so I "assigned" him a duty. I had been involved with two buildings in his type of business earlier. I explained in detail how these two buildings were different from each other and were still used for the same purpose, and that I wanted him to see them. One seemed to be exactly what he had in mind. I gave him directions, and it was only a 45-minute drive from his home, so he planned to take a look on the following Sunday. The next Tuesday I stopped off to see what he thought about the store. Well, he hadn't gone over to look at it.

If you were going to build something, and you found out where you could see what you had in mind, would you go look at it if you were really interested? You bet you would! Owner X wasn't interested enough to go look; so he must not have been really interested at all. I dropped the lead as I would a hot potato.

The interest area is also a gray zone. It's quite hard sometimes to work up a readout on the lead; and, believe me, nothing is guaranteed in construction selling. I've found from experience that you will seldom disqualify a lead from lack of interest. I think every businessperson would like to talk about a new store or expanding the warehouse. But it's important not to take interest for granted; when you do, your chances of being used will increase. This is definitely not good for the construction sales agent.

I follow a general rule on this subject and think about it enough to keep my perspective: Do all you can to maximize the number of good prospects, and do all you can to minimize the manipulators. You will be manipulated! It really will happen, and it will probably upset you, but don't let this unpleasant side of construction selling turn you off. It comes with the territory because you're dealing with individuals before they are completely ready to build. The whole idea of qualifying the lead is to cut down on the instances of just being used. Remember, there is nothing at all that says a good prime prospect will not nail you to the wall.

When I first started selling construction, I was an eager beaver and gave the lead or prospect everything he or she asked for, most of the time with no questions asked. My "being used" percentage was pretty high. As I gained more experience, this percentage steadily dropped. Another factor has had a small part in this percentage drop: age. I honestly can't say much for becoming older except that it seems to help your business image.

There are two ways of being used. The first one is almost impossible to avoid. The lead checks out and you start selling; you take all the information, do your homework, and put together a price. The prospect takes one look at the numbers, has a fit, and discards the entire project. You know there's no way to cut costs and still meet the requirements, and the prospect won't compromise; so there you are. All that work for nothing, but—and this is an important but—neither one of you knew this in advance. So I suggest you make it a point in such cases to part friends because your defunct prospect may decide to make some changes later. Better yet, the construction salesperson should keep in touch. A phone call every so often will do the job.

In the second type of manipulation, the prospect is using you to obtain every possible scrap of information, drawings, etc. The prospect knows all the while that he or she doesn't plan to build now, but the prospect doesn't hesitate to take advantage of you. Sometimes a person will do this and see absolutely nothing wrong with it. After all, you did call on the prospect and ask for the business; and since you offered, the prospect intends to get all he or she can. This type of person, I've found out, will let you run your legs off but will do nothing personally. Assigning this prospect a duty to perform will flush him or her out every time. The duty will not be performed, and the excuses you'll hear might be somewhat entertaining.

Sometimes the interest factor can be combined with other areas of qualification. For example, the interest emphasis may be put on the site so that you're checking two qualifiers at the same time, which will give the sales agent a good handle on the lead. There are other ways to combine qualifiers, all with the interest factor. Again, for example, if I get the feeling the lead is a blowhard and he or she tells me money is no problem, I respond immediately with how glad I am to hear that; and I say that since he or she is a good businessperson, I know the lead will want any general contractor bonded who does work for him or her. That being the case, I'd appreciate a copy of the loan commitment for my bonding company so that there will be no doubt about the money back at the home office of the bond company. If the lead is really using you, he or she probably hasn't even talked to a banker, much less obtained a loan takeout letter. At this point the lead starts backing away. His or her hand has been called with some heavy stuff. The lead won't like hearing about home office bonding people being interested in him or her. The lead will realize he or she is dealing with a professional, not some rinky dink person to mentally manhandle. I've used this interest "dig" quite a few times, and it works every time. I've been surprised also: Twice the lead handed over a copy of his takeout.

I'm sure with time the new construction sales agent will be able to feel comfortable in this area and will have confidence in her or his decisions. You can't be right all the time, but when the percentages are totaled, you'll be ahead of the game. Remember, you're talking about business with owner X, so don't be afraid to ask questions about the business and the planned facility. You can never have too much information.

I know I've painted a grim picture, and it seems you'll be out there on the streets, looking over your shoulder all the time. Not so! Everything has a price, and this is the price you pay in selling construction. But look at the other side of the coin. There will be many profitable projects done with first-rate individuals; some will even end up friends. I sold three 18,000 ft^2 roller-skating rinks to one owner. This owner is now a good friend, and every year my daughter, while in high school, received a free pass at any of his rinks. (Of course, my friend is also a good businessman, since every time my daughter went skating, she managed to fill the car with skaters who had to pay!) So don't be downhearted; there are plenty of good times in selling construction.

To sum up the interest factor, if owner X doesn't display any, then move on to the next lead; without interest, there's no sale.

Talking to the Decision Makers

More time has been wasted on talking to and working on the wrong person than on all the rest of the qualifier problems put together. The name of the game is selling percentages, but here the construction salesperson has a mental block, especially the new one.

I can understand perfectly how you can get caught up and carried along with the momentum of putting a project together. You make a call. The owner is easy to talk with and is planning on a new warehouse. You ask about the site, and the owner takes you out back and shows you the building location. You ask about the money, at the same time knowing the firm enjoys an excellent credit rating. You establish the reason for building—taking on two new lines of equipment. There's interest shown. When all these subjects click, the good construction sales representative is like a racehorse in the starting gate. And when the prospect asks for a proposal as soon as possible because he or she needs the new building by such and such a date, then look out—our salesperson is cranking up! It's exciting to put a beautiful deal together and very easy to become intoxicated with it. So our eager sales representative puts together a proposal and presents it to the owner. Owner X looks over the information and casually says, "Looks okay. The price is within our budget also." Our sales agent is ready to whip out a pen so the owner can sign. Then owner X drops a bomb: "I'll get this right off to Ms. So and So; she's the one in our main office who'll make the decision."

At this point, the sales agent has just said goodbye to a whole lot of work, time, and money. This can happen to you if you're not careful. I know from painful past experience.

There will be exceptions. For example, the branch manager is a friend and he asks you for a proposal. You don't mind working one up because you feel you'll have a friend to help you. These exceptions will be few and far between. As for the rest of the leads, keep both eyes open, and know just what it is you're getting into. Beware of branch managers; they are the best manipulators the construction sales agent can encounter. I mean nothing personal; I have some close friends who are branch managers of out-of-town firms. It's in their nature to use the construction sales agent to help them. They all want more space, office, storage, etc., and they persistently hound the home office. When the sales agent walks in and offers to give the branch manager some more ammunition to send to the home office, the manager takes her or him up on the spot. And I would do the same thing in the branch manager's place, I'm sure.

The best defense is to find out right away to whom you are talking. Use a little finesse. I've found that if I ask someone who obviously is not in charge who is, and then the owner or manager is pointed out, this helps me through what could become an awkward situation. Anyway, be certain to whom you're offering your services.

When you determine that owner X is indeed the manager, go right on with the qualifying and requirements as if the lead were a prospect. Tell owner X you'll be glad to work up a proposal; then ask when the person from the home office will be in town—it would help to know this so that you could schedule your time. Also be quick to add that if time is very important, you'll be glad to contact the home office. You're telling owner X in a nice way that you're not going to do business with her or him. You expect to deal with the decision maker.

I have never sold a job where I couldn't talk with the decision maker. I have sold several very clean, fast-to-build, and profitable jobs to out-of-town companies in which I could make my sales presentation to the person who could make the decisions. I'm speaking from experience; I was taken to the cleaners so many times that even my wife told me to wise up. Now I don't make proposals to managers unless it's immediately clear that I am talking to the decision maker.

You may feel that you're walking away from some business, but you're really not. If you are lucky and get such a job, there's never enough money in it to begin to pay you back for all the work you put into the ones you missed. Percentages again—that's all selling is; and the successful salesperson knows how to play them to advantage.

I know you'll uncover a lead who turns out to be a supergood prospect, with the only problem being that you're dealing with a manager. It looks good, and you want to give it a shot. Go ahead, that's fine—but don't let it take away anything from a better prospect, even if the prospect doesn't offer the same inducement as the manager. Look at the situation as investing some money in a high-risk operation. Invest only what you can afford to lose. Play the percentages; better yet, believe in them. A good friend of mine, who's an air force pilot, told me the hardest thing he had to learn about flying was to look at his instruments and believe them, no matter what he thought personally. The same goes for the construction salesperson. Learn to read the signals and then believe them.

There's yet another pitfall to watch out for: the parent-and-daughter, son team. It's hard to determine whom to talk to. If you start out with the parent, chances are you'll be okay; but if the daughter or son is the one wanting all the information, be careful. The child has to be considered in the owner category, but chances are the child won't be the one to make the decisions. Also the child will be eager to do something—he or she will be interested in making the business grow and so will jump right on the services you can provide. Then the parent will say no for some reason or other. Be prepared also to say everything twice with the parent-and-child lead; for some strange reason you can never see them both together—one always has to watch the store. Just try to find out all you can about the parent-and-child lead. It's important to talk to the decision maker without making the other one (usually the child) upset because she or he is not the one running the show. You need to be an excellent diplomat in this situation.

To sum up, if you cannot talk to the person who will say yes or no and sign the check, you're wasting your time. This qualifier is just as important as any of the rest I've explained, maybe even more so; for it is the one that's the easiest to overlook and/or want to overlook, so that the salesperson gets sucked in.

Personal Problems the Lead Might Have

This subject is not a qualifier in the same sense as the preceding ones, but knowing about the lead's personal problems will help the construction salesperson make better use of her or his time. Note that I'm not talking about personal problems in the sense that immediately comes to mind; I'm talking about the personal problem that is talked about in public.

Here's a good example. I worked with a good prospect who planned to build a 20,000 ft^2 metal building, 34 ft high, to be used for boat storage. In the course of working out the requirements, the owner mentioned that he wouldn't need the information for 3 or 4 weeks. The owner was due to have an operation and would be out for almost a month. With that information I was able to work my schedule around the proposal to my advantage.

Most of the personal problems that I've had to contend with have involved sickness, and I'm sure your experience will be the same. Sometimes the construction sales agent picks up a piece of information that really can be a semiqualifier. Always ask the lead what the time schedule is; usually here is where the problems will come to light. At the same time, if you hear something about owner X having trouble trying to buy out a partner, and that the owner will not be free to pursue the project until that's taken care of, which should be in such and such a time frame, be a little cautious. Don't jump up and run, but be wary.

Remember to be tuned in to anything that will help you. Pay attention when owner X says something about personal problems. You may even be subjected to hearing about an upcoming divorce or whatever. If so, listen; you can never have enough information.

A lead is nothing more than a name until you qualify that lead into a good prospect. A good prospect is one who passes all the qualifiers and plans to build. The qualifiers that you will use are site, finances, reason for building, interest, talking to the person in charge, and personal problems.

It's imperative that the construction salesperson understand how these qualifiers are used and why. Their use should become second nature, and with experience it will be possible to qualify a lead in a general way without calling on the person.

When a friend passes on a lead to you, ask if the owner has a site and money. Your contact may have the answer and may save you some time. But don't depend on it too much. It pays to contact the lead even though you know he or she doesn't have a site, to show you're interested so that you can keep your name up front. Also, you may be able to recommend a solution to the problem. As a matter of fact, a solution for the problem would be an excellent reason for contacting the lead.

Asking questions from every source, digging out details, and putting all the information together are going to be second nature to the really good construction sales agent. And I'm sure you are similar to me: You'll have to develop these techniques. No one is born with them. Some people seem to be more people-oriented than others, which helps, but it's not mandatory. Hard work is!

Learn how to qualify that lead; if you don't, all you'll be doing is spinning your wheels. Nothing else is as disheartening as working and getting nothing to show for it. This chapter will help you minimize such wasted effort; it will also help you improve your contract-signing percentage by allowing you to work prospects who plan to build. All you have to do is to convince them to let you do the job.

NOTES

NOTES

4

Advantages of Using a Metal Building

In the commercial and industrial construction business, the metal building gives the owner a number of pluses that will be looked into with a great deal of interest. The sales agent will find these areas very helpful when presenting to owners information supporting the choice of a metal building over standard construction.

Let me repeat a statement made to me by an owner after I finished going over advantages of a metal building. He agreed that for his business the metal building was by far the best way to go, since he was a distributor of items that came in on large trucks and he sent to his customer the items in his local delivery trucks. He said, "A metal building is the best way to get a big box to make money." We both laughed, but it's a fact, and I have used his comment many, many times when going over the advantages with prospects. It's also fun to hear their chuckle or note their smile when they realize you're a nice guy with a sense of humor, which helps the salesperson-prospect dynamic.

Cost

The first advantage lies in the building cost, and since size and use will vary, I'm going to cover this in a very general overview, looking at the building simply as an empty building.

When you compare the smaller buildings, say 2000 up to 5000 ft^2, the metal building can come in more cheaply if the outside is kept very plain with no or very little material used to improve the architectural look. So, depending on the use which affects the looks, the cost will vary; but when compared to standard construction, the metal building should be cheaper.

The real cost advantage comes into play when the building area is greater than 10,000 ft^2. The larger metal building really becomes cost-effective and makes the bottom line very attractive to the owner.

So if you are selling a 5000 ft^2 building for a paint supplier to store materials, address cost, but put greater emphasis on the other advantages.

When you are working on a large building, say, 20,000 ft^2, the cost should be lower, so give the cost advantage a lot of attention, because you should be talking about a nice savings for the owner, and this will really hold his or her interest. After all, the only reason the owner wants this building, and you want to build it, is money; so when you can use it to your advantage, do so.

Faster Construction

Time is the one element that everybody knows the building contractor has no clue about. So when you can talk to the prospect about a faster construction time, it makes you look good. Outline to the owner that the metal building manufacturers run on a schedule, and that the contractor has to arrange everything around this schedule.

I'm truly amazed when I see a metal building that's been unloaded on a site where no other construction work has been done. Then the building sits on the ground for months in the weather, and I bet the owner makes a lot of phone calls asking when something is going to happen.

So I use the building delivery schedule to my advantage by making it the focus point for everything. For example, usually it will take 6 to 10 weeks to have the metal building delivered after it is ordered. I use this time to have the slab, anchor bolts, and foundation work completed; so when the building comes in, it's unloaded, and erection starts that day, if possible, or the next day for sure.

I recently had an owner and his wife show up the afternoon of the delivery day with lawn chairs to sit and watch the erection start up. He's since put me on to two good prospects.

Now when it will take more time to start construction because the permit process has added time or it's simply bad weather, then I work with my metal building manufacturer and push the delivery date back to a point that will work with my schedule. That way I am still in charge.

Given faster construction advantage, references are very helpful, so give the prospect a couple of names to check out. At the same time, I have invited the prospect to visit a job I have under construction to see in person what I'm talking about.

Metal building can be faster construction, but it's up to the contractor to use this advantage.

Flexible Building Uses

While cost and faster construction are good advantages to help sell the building, metal building flexibility is by far the area that really piques the owner's interest. This is because, as the business grows or changes, the owner can easily do things to a building to keep the business moving ahead.

The first point I bring up with the owner is the fact that the roof system is not supported with the walls. (Now I know it can be in special situations, but for the area I work in, the standard metal building is fine.) Then I explain exactly what I am talking about, and when possible, I go with the owner back in the shop or storage area and point out the block walls that hold up the roof.

Next I tell her or him what a big deal it is to cut an overhead door opening in the wall because of the load it carries. Then I explain that steel columns hold up the metal building roof system, and the wall area between the columns does not. So if a walk door, an overhead door, a set of windows, or an opening into an addition is needed, it is easily placed in the wall.

The owner will pick up on this immediately as a benefit when growth and changes require greater convenience in the layout and use of the building.

The next plus of the metal building is the ability to give clear span where needed so that roof support columns are not in the way. A lot of owners are really pleased when they note the large clear area that's possible. The smaller buildings that are from, say, 25 to 60 ft wide can have clear span with no trouble. Now when you get into wider buildings, for the clear span, no columns are still available, but the cost starts to go up. It's the fact the larger clear span is available, and that is what's needed to meet the planned use of the building that makes the added cost acceptable.

A good example is a skating rink. Can you imagine having to dodge steel columns while skating, be it on ice or roller blades? Right now I am talking to the Parks and Recreation Department in the town where I live about a metal building to have indoor games such as volleyball, basketball, and other types of gymnastic activities. The building is 100 ft by 180 ft with 30 ft of eave height and a 1-in-12 roof slope. They know the 100-ft clear span will cost more, but it's what is needed; and as somebody at city hall noted, I can help pay the extra cost since I am a tax-paying resident.

There are other special areas that require or want clear span space. I know of an example of one building being used by two special areas. A skating rink I was involved with several years ago has been sold to a church and is now the main assembly area. So follow up and become familiar with the areas that can use the metal building clear-span availability in your work area.

While most businesses can put up with columns because they have permanent equipment to place, and the layout can be arranged around the columns, there will still be the occasion on which clear span will be very helpful in a special section of the building. I have a good example. I recently was contacted by a large moving and storage firm to repair a column located in its loading area (Fig. 9). This area is where large boxes are moved off and on the trucks by forklifts, and this area has two roof system columns that have forklifts moving around all day long. This is a perfect place to have clear span while the rest of the storage building can have columns. So when you are telling the prospect about the clear-span availability, stress the fact certain areas can be addressed and not the entire building.

The fact that columns support the roof system means no interior construction is used to help support the roof system. Let me give an example of how I use this fact when selling a job. The prospect wants a building 60 ft wide, 150 ft long, and 20 ft high; 2000 ft^2 of offices and 7000 ft^2 of warehouse space. Now the 20-ft-high walls are needed in the warehouse, so materials can be stacked. However, the office area will do fine with 10- or 12-ft sidewalls, so the usual way to do this is to put the office in front, attached to the higher warehouse. But if you can convince the prospect that the whole building should be 20 ft high, you stand to make a little

Figure 9. Two views of a damaged column.

Figure 9. (*Continued*)

more money, and the construction process will be simpler to do. Plus you will have less chance of callbacks after the building is in use due to leaks where the buildings come together.

This is how I approach the prospect. I use the fact that nothing inside helps to hold the roof, so offices at any time, for any reason, can be changed or added because the walls are nothing more than partitions.

Then I present the prospect with an idea that will be most useful and helpful in the future. As the business grows and the prospect needs more office space, new offices can be added on top of existing ones if 20-ft-high walls are used for the whole building.

So by building upward the offices are not expanded into the warehouse space. Also, while extra offices are not needed, the space on top can be used for storage. I have sold this approach a number of times. Both the owner and I come out ahead.

This approach also has another aspect that the owner will find very useful if she or he plans to rent out the building. The owner can give the tenant what is needed inside the building, and down the road when tenants change, the owner can make whatever adjustments are necessary for the new tenant. I have done several buildings for real estate people who have found this feature very convenient. Plus, often I am called on to make the changes.

Let's look at another area of metal building flexibility—shape! The building does *not* have to be a square or a rectangle. The metal box can be shaped to any convenient angle necessary to get the job done. (See Figs. 10 and 11.)

The next place to take the prospect is under a canopy. If for this particular type of business a canopy will be very useful, explain how easily it blends in with the metal building. A good point that I stress is that often a canopy does not need columns, which makes the loading area more convenient. (See Fig. 12.) Then note how flexible the canopy is since it can be easily expanded, and in some cases wall panels are added to improve the use. I have found pictures very handy when talking about canopies.

Present the metal building as a freestanding box in which most anything can be done inside or outside at any time. It is a very good sales angle from your side, and the prospect will find it most interesting.

Now that you have the prospect thinking about future expansion, let's get into the next area to follow up with, the expandable end wall. Again the contractor will get some more money, but it's money well spent by the owner, and that cost will be made up when the building is expanded. Plus, you have future work to follow up on as time slips by.

While we are on future expansion, let me bring up a special area that I tell the prospect I will do at no extra charge. All the owner has to do is to give some thought to how the building will be used, and this gets the prospect thinking about the building he or she wants to build. I'm talking

Figure 10. Two views of a building showing metal building flexibilty

about overhead doors. Let's say the prospect's building will have two overhead doors to start with; but when the business grows, as the prospect expects, then she or he will need at least two more overhead doors located at points *A* and *B*. I then explain that the walls don't hold up the roof. The area between columns is called a *bay*, and although there may be four, five, or six bays on each sidewall, one or two of them will be used for cross-bracing. So when the metal building manufacturer fabricates the building, the future overhead doors are noted and their bays are not used for cross-bracing.

Figure 11. A building over a loading area that stays in use is very flexible.

Figure 12. A canopy over a loading area.

Then I explain in detail how anchor bolts will be placed and the concrete finished to receive a future overhead door; and since the wall is not held up by the slab, the bolts and sloped concrete cause no problems. (See Fig. 13.) While going through this, I tell the prospect that it's not too hard to install the overhead door, and there will be very little inconvenience for the business while it's being done.

Another area I get into with the prospect makes me look good. I ask the prospect how the overhead doors will be used. Will only trucks or cars be coming or going, or will there be forklifts and handcarts using the doors as well? Often I have already seen how the doors are being used by going back in the shop, but I still bring up the question.

You see a lot of commercial overhead doors that have a metal flange where the doors sit on the floor. (See Fig. 14.) While this arrangement keeps the rainwater out of the building, it gives forklifts and handcarts a bump to get over which can be a problem sometimes. My solution is to slope the entryway a little farther into the building, and not to use the metal bump. (See Fig. 15.) This lets the forklifts and handcarts come and go easily, and the rainwater runs down to the outside.

I recently put up a 10,000 ft^2 addition for a seafood processor, and the owner liked this idea, but I really enjoyed the thanks the employees gave me because most of the seafood, as it was processed, was moved around in large handcarts.

While we are discussing overhead doors, let's look at protecting them. These doors are used mainly by trucks of all kinds, which means they will

Figure 13. Future overhead door location.

Figure 14. Metal flange in door entry.

Figure 15. Doorway without a metal flange.

get hit so let's put something in the truck's way—a door bumper. Now we are used to seeing the bumper on the outside, but how about the inside? (See picture 16.) Having protection from both directions can really aid in maintaining the door.

Another good subject to explain is metal building maintenance and repairs which most of the time can be very simple, again since the sidewalls don't hold up the roof. Often metal buildings will have a lot of traffic from both trucks and cars around them, so the buildings get bumped into. To replace corner trim or some wall sheets is an easy job compared to repairing a block wall.

Our last area of metal building flexibility is quite unique. (See Fig. 17.) What we have here is a metal building placed on a block wall building that was not designed for a second story. This building is a municipal building used by the town where I live.

The idea of using the metal building as a method of enlarging an existing building by going upward is fascinating to consider and a very good selling point where it can be used.

Metal building flexibility will give you a positive picture to present to prospects and will get them thinking about the business, and the changes they will have to consider in the future. The other plus is that you, the

Figure 16. A door bumper can protect a door wall inside and outside.

Figure 17. A metal building on a block building.

Figure 17. (*Continued*)

contractor, come on to the prospect as a businessman, and not someone simply wanting to pour concrete and set steel.

In App. E, Framing Systems, you can see some examples of frames, showing the flexibility with which a metal building can be constructed.

The bottom line here is that all these advantages of a metal building will be used in the selling process to help the prospect decide to take advantage of a metal building. Then the next thing you must do is to convince the prospect that you are the contractor to do the work.

While this chapter is aimed at the prospect who is not familiar with a metal building, it may come in handy when you are talking with a prospect who already has a metal building that was constructed by someone else, and some of the advantages were not taken into account. So even though the prospect is sold on using a metal building, you still make yourself look good by bringing these ideas into the picture.

I know I have covered these advantages in a very general way, and I have made it sound easy to do. So now I am going to muddy the water by bringing up building codes.

As you know, building codes differ according to the building materials, use of the building, and the building location. So you will have to consult the building codes in your construction area and make them fit the use of the building. Tell the prospect that the codes will have to be worked in. But I found out it helps not to go into a lot of details with the prospect because this is one area that most people are not even vaguely familiar with.

So the prospect is sold on a metal building. Now let's move along with selling the prospect on your doing the job.

NOTES

NOTES

5

The Art of Selling to the Prospect

Selling is an art, not a pure science. It is dealing with people; thus, there are no ironclad rules. The rules change with the individual, so the person practicing the art of selling has to be able to adapt to the situation.

Selling is simply one person's setting out to convince someone else to buy his or her wares. Everyone who is in the process of earning a living sells. It can be the employee behind the counter of the local hardware store selling tangible goods, or it can be the insurance agent selling intangibles. The secretary is selling skills. The doctor thumping your chest is selling knowledge. The lawyer sitting quietly, listening to the details of a potential lawsuit, is selling services. When the general contractor puts in a bid, the idea is to sell the ability to perform the work, at the lowest price. This is selling as thought of in its broadest sense. Many people do not think of themselves as salespeople pushing their goods and services, because this idea may not be an accepted part of their way of earning a living. The term that I like to use for the selling of those people who sell but who don't recognize the fact that they do, is *passive selling.* Office workers, military personnel, and factory employees are all passive salespeople, making up by far the bulk of those in the job market.

The other side of the coin is represented by the person who accepts the fact that selling is one of the most important business functions that he or she can perform—the aggressive salesperson. And I don't mean the "foot-in-the-door" type of aggressiveness, but rather the "go to the customer and ask for the business" type. Many people practice the art of aggressive selling. You see them everyday: the building supply sales agent who calls on customers regularly, car dealers, shopping center merchants, and general contractors—they're all out in the marketplace practicing aggressive selling. They are trying to get the customer to spend money with them. The result of this effort is the greatest success story the world has ever known: free enterprise.

The goal is to help the metal building contractor bring his or her aggressive salesmanship to its highest peak by going out in the marketplace and successfully selling his or her skills, knowledge, and abilities to an individual in need of those services. Now this is an entirely different type of selling from what the metal building contractor is used to. Most of her or his background and experience has been in bidding for business; but now we are talking about going and asking for the chance to build a new building. The person who does this, be it the owner or an employee, will have to become a true, honest-to-goodness salesperson in every sense of the word.

On and Off Buttons

The first point a salesperson must understand is that people have on and off buttons, and knowing how to push the on button is the first step in ending up with a signed contract. At the start you are not worrying as much about the on button as you are about taking pains not to push the off button. For example, I'm sure at some time you have walked into a clothing store, say, looking for a new sport coat. You quickly find the sport coat section, and about that time the clerk comes up and asks to help. You tell the clerk that you'd like to browse for a bit; the clerk proceeds to ignore your request and hovers at your elbow, telling you how this coat or that coat is made for you. After a few minutes of the pushy selling approach, you leave—not because there was a poor selection of sport coats but because the clerk pushed your off button.

Think back over some of the times a salesperson has turned you off because of the manner of the approach. Now think about the good experiences, when you have said to yourself, "I like that person, and she knows her business." Chances are, the differences lie in the sales personalities of the individuals with whom you are dealing. The way a person comes across to you at the start creates an impression that is almost always permanent.

When you call on a prospect, no matter whether it's a cold call or one that was recommended, you want to create a good impression. You want to come across as a nice person, thoroughly grounded in your business, who thought enough of owner X's business to come in and ask for it.

Whenever possible, call for an appointment. This phone call does two things for you: It shows owner X that you run a businesslike firm, which is helpful in having X start to form the correct impression of you; and the call lets you find out if the owner does indeed plan to build, which helps you make better use of your time.

I recommend that when calling a perfect stranger on the phone for the first time, you state your name and your company name and then ask if that person has a few moments to talk to you. One of the rudest assumptions is that the other person has time to talk just because he or she is on the other end of the telephone line. So ask if it's convenient; and if not, say you'll be glad to call back. And ninety-nine times out of a hundred, owner X will tell you to go right ahead. So you've cracked the door, and owner X hopefully will start to form that favorable impression of you since, by asking for a moment's time, you have in effect said, "I know you're busy, and we're having this conversation because you want to." You have put the owner in charge, and everybody likes to be there. After the preliminaries, get to the reason for the call and then say goodbye.

Many times it's not practical to phone ahead, for example, when calling on a likely prospect whom you know nothing about. But it makes no difference whether you have an appointment or are making a cold call. The idea is to not push the off button.

When you have an appointment, be on time! If for some reason you cannot be punctual, call as soon as you know and do give a reason. The owner is in business and has problems, so the owner will understand when you say a truck backed into a metal building wall and you have to be there.

Now remember, the only goal you have set for yourself so far with owner X is not to push her or his off button. You are in essence simply selling yourself, and you are the first thing the owner has to buy, long before the contract is signed.

You arrive at owner X's place of business, report to the secretary that you have an appointment, and finally meet owner X face to face. But hold it a minute—suppose owner X runs an auto parts store and there are five customers waiting. Then you stand to one side out of the way until owner X has everything under control; you wait until he or she can see you. Then the owner may leave you to wait on a customer. The owner will always put you second to the customer if it is a walk-in type of retail business—a fact you'll have to live with. The point is, make yourself fit into owner X's scheme. Sometimes it's not easy, but it is necessary.

Regardless of what you have done to meet owner X, you are finally there; and, as is customary, the two of you shake hands. Now ask yourself: What did owner X think of my handshake? This is important because I firmly believe more off buttons have been pushed by "limp fish" handshakes than by all the other reasons combined. If you don't know what kind of handshake you give, then I seriously suggest that you find out. Nothing makes a better impression than a brief, firm handshake.

Now that the meeting is underway, you, the sales representative, will have to do the initial talking to get the interview off the ground. After all, you've come to owner X, and she or he wants to know the details of why you are there. As the meeting progresses, the good salesperson starts to control the conversation. Note that I said *control,* not dominate. People seem to have the preconceived notion that the hotshot salesperson is a smooth, fast talker who never gives the buyer a chance to say a word. Not so!

The really successful salespeople are the best listeners. The first item of importance in the interview is to learn. You need to find out what the prospect's plans are: Does he or she have a site, has financing been obtained, and can you help in any way? Are there any special problems that must be looked into? Learn all you can; you can never have too much information about a prospect. All this evolves by asking the correct questions—and then sitting back and listening. The impression made is one of trying to be helpful, not one of telling the prospect that you have just the thing. Rather, you're asking what the requirements are and how you can help by using those requirements.

By the way, never, never tell owner X how to run the business; that's a sure way to push his or her off button.

There are three particular off buttons that I feel strongly about: the beeper, the car phone, and the office phone. Nothing is as distracting as a beeper's going off right in the middle of a meeting with a prospect. And I'm not the only one who feels that way. Ask around, and you'll discover (as I did) most people don't like it. So if you use a beeper, take it off and leave it in your car. Even if it's turned off but still on your belt, the prospect doesn't know this and may be waiting for it to beep.

If you have a car phone and your prospect is in the car to see some examples of your work, turn it off. You have a captive audience, so take advantage of it. Don't think you'll impress the prospect with how important you are, as you tool around town, because the office simply has to get hold of you. Forget it—the prospect will not be impressed. Instead the prospect will think about what's happening back at the store and then find some reason to cut short the trip.

If the prospect comes to your office for a meeting, hold your calls. If no one is up front, turn on the answering machine.

It's hard enough to get and keep the prospect's interest, so when you get the opportunity, do everything you can to improve your chances.

Low Profile and Soft Sell

At this point what the salesperson is doing is maintaining a low-profile, soft-sell attitude that does not put off the prospect. The salesperson is selling her- or himself and making owner X start to think about doing business with her or him.

Low-profile, soft sell is the exact opposite of the obnoxious, loud-mouth, contract-waving, slick talker. Low-profile, soft sell means doing your selling in a courteous, respectful, businesslike manner. You never presume to tell the prospect what he or she needs until you have all the facts and the prospect asks for your opinion. You are helpful, and you don't get bent out of shape when you are inconvenienced by the prospect because of the pressures and responsibilities of running his or her business. You understand the prospect's particular problems and are able to fit yourself into the prospect's picture. You sell to the prospect in the same way you would want to be sold to if the roles were reversed.

Many useful things can be found out during the first meetings with the prospect, but most are technical and will be covered in detail in later chapters. Right now we are interested in the general, overall selling picture and how it works for the metal building sales agent.

Metal building salespeople have one important thing working for them that puts them right at the top of this profession: They are selling goods that are not purchased unless needed. They never push something off on the customer. When they sell a building, it's needed. This gives construction salespeople a mental attitude about their work that allows them to think of themselves as truly professional in their business community. The reason is simple. Owner X might buy some $10 gimmick for his or her car even though the owner might not need it, but the owner will never buy that 5000 ft^2 storage building unless she or he really needs it, no matter how good the salesperson is. Construction salespeople provide the same services as any other professional; they fill the needs of the buyer when the buyer is in the market for their special services.

So far our selling attitude has been centered on the idea of low-profile, soft sell, but the trick is to achieve this and remain aggressive. The sales agent must know just how far to carry a subject, when not to pursue a certain avenue; and this knowledge can only come from experience. Every good salesperson has to work out her or his own system.

The main way to remain aggressive while practicing low-profile, soft sell is to show interest. There are many ways to show interest during the selling period of contract negotiations, and those ways will be discussed in detail in later chapters.

Time Interval

There is an amazing fact concerning businesspeople that may or may not be known to you: Owners will finally decide to build that new facility only after they are forced to face the hard fact that they must have a new building. They will hang onto their old place for dear life until the very last minute.

For a businessperson the prime mover is always money; in every decision it is the predominant factor. Why should owner X build the new warehouse and double the monthly mortgage? The owner will build when forced out by redevelopment, or if a new highway is going through the loading dock; business expansion is always a long and agonizing undertaking. The reason I mention this now is to point out that during the selling process it may take months or even years, in extreme cases, for the owner to be ready to sign up and get the ball moving. Construction salespeople must accept this fact. Very few sales are made quickly; most take between 2 and 6 months. And if you are doing your job effectively and efficiently, this is the ideal time sequence. The idea is to make contact with the prospect in the initial planning stage and carry through to contract signing, then on to the building stage. Since most businesspeople move cautiously, it will take time to do the selling job right.

A quick project, that is, one where the prospect has done a great deal of preliminary work and wants to get started right away, is a project that is already out on the street. The end result is one that boils down to the low price, and that is what you are trying to get away from by selling metal building projects.

Obviously, then, for the first months you or your sales agent is operating, there will probably be no contracts coming in. It will take time to get prospects lined up so that several are being worked on at various stages; then before long, the repeating process will take over, in which jobs are going to contract as new prospects are added. Just bear in mind that if you end up with some fast business, it will be due more to luck than to anything else. So don't be discouraged when you start selling and get the idea that you are doing nothing but spinning your wheels.

Empathy

A good construction salesperson should be able to practice empathy. Empathy is the ability to put yourself in someone else's place and see things from that point of view. From the sales agent's side, it may look as if the owner has no choice but to build, so you get in gear and go after the job. Then owner X tells you why he or she is not going to build; and unless you can change places with the owner, it may be hard to understand.

Empathy can also make your job easier. For example, you get the feeling that a prospect is putting you off; she or he is not doing enough to push the project along, even though both of you know the prospect plans to build. Looking at the situation from the prospect's viewpoint may show you that she or he really does not have a time problem and is not pushing you around.

So far we have talked about the fundamentals with which the construction sales representative should be comfortable: practicing low-profile, soft selling and yet remaining aggressive; thinking about how to push the on buttons and how not to push the off buttons; being available at the convenience of the prospect; learning to listen; impressing the prospect with your interest in the business; accepting the sometimes excessively long time intervals; and practicing empathy.

All these suggestions are only guidelines, not ironclad rules. Selling is an art, not a science. Every individual who undertakes to sell construction projects aggressively is going to have to work out what is best for him or her. However, here are some important areas that should be given careful consideration when you are eyeball-to-eyeball with owner X.

As you read on, I will refer to these basic guidelines with specific details to help you get a better grasp of them.

Understanding the Business Prospect

At this point I would like to consider the prospects whom the construction sales agent will be calling on by business category. There are three main areas: retail, wholesale and industrial, and professional.

Retail

The first thing that you will notice about the retail segment of the business community is they think small (I don't mean this in a derogatory manner at all). The retail merchant for the most part sells items that cost from pennies up to several hundred dollars; he or she is used to dealing with small sums of money. At the same time, her or his business outlook is shaped on a day-to-day basis.

Prove it to yourself. The next time you visit your local hardware store or drugstore, ask the owner or manager how business is. The answer will probably run along these lines: "Not too bad today, but yesterday was really off." To the construction person who thinks in terms of good and bad years, this day-to-day existence is a little difficult to fathom, and sometimes hard to deal with.

When the retail merchant has a bad day, he or she has a hard time spending money; and believe me, one bad day can ruin the whole deal. For example, I was working with a retail merchant who had been leasing for years, and he had finally reached the point where he was ready to take the plunge and build his own place. Business was good, and after 3 months of preliminary work, he told me to draw up the contract. The very next afternoon I stopped by for his signature and was startled when he told me he wanted to hold off for a while. The man sold toys, and that particular day was awful. Based on that day he decided to postpone the project until after he saw how his Christmas business would be. Well, I ask you, if a toy merchant can't look forward to and expect good business during the Christmas season, just when does he expect it? The man never built; there was always some reason not to.

Timing is very important when you are calling on and working with the retail businessperson, so make it a point to find out the best time to make contact. She or he will appreciate knowing you don't plan to take up busy times.

It is maddening at times to try to carry on a meaningful conversation with the retail merchant who is watching the sales area; and even if the merchant uses the office or walks back to the stockroom, she or he is still subject to interruptions. Remember, as much as the retailer may want to talk to you, customers will always come first.

A method I have found to be most useful is to get owner X away from his or her place. Lunch is a fairly productive solution, but there are always distractions. The best way to get owner X away from his or her clerks and the phone is to take the owner to look at a building that is similar to what she or he has in mind. The prospect then becomes a captive audience, and the salesperson can accomplish a great deal in a short time. Another good method is to arrange a meeting after hours, say, in the evening or over the weekend. The important point is that the retail merchant is at times difficult to talk with, and the good construction sales agent has to be able to maneuver the prospect into a meeting that will pay dividends.

Let's go back to the money angle. As already stated, most of the time retail merchants deal in small sums of money, especially when compared to contractors. Say $300,000 to the average retail merchant, and notice the color slowly drain from his or her face and a slight glaze cover the eyes, whereas the general contractor is used to running large sums of money through the books. The average contractor thinks no more of $200,000 than of $2 million. The trick in the construction industry is to have some left after running these large amounts through the books. Right?

The retail merchant who is in the market for a new facility will generally have an idea of what the new building should cost. The figures will probably be at least 10 years old. So when you give some budget numbers, owner X will gasp and say, "I had no idea of spending that much; it's out of the question. I'll just stay put for a while."

At this point the construction salesperson has to educate the prospect and slowly get her or him used to the numbers. This takes time and persistence. Remember, this money to the contractor is really not very much, but to the average retail merchant it's probably the largest money transaction he or she will ever undertake. The merchant is reluctant and apprehensive. The sales agent must understand this, accept the fact, and work toward making the prospect comfortable with the figures.

Sometimes the retail merchant does not have the financial background concerning large sums of money that you might expect. He or she has no idea at all about the details of pulling together a financial package to pay for the new facility. Here the construction sales agent can be most helpful and, at the same time, make her- or himself more important to owner X—which is the name of the game at the selling stage. To sum up, the construction salesperson must keep in mind when working with the retail businessperson that the prospect is the exact opposite of the construction person in the manner that business is viewed and conducted.

Wholesale and Industrial

Wholesale and industrial business is as different from retail business as day is from night. This is the segment of the business community where you as a metal building contractor will be able to establish a very good rapport.

Like contractors, wholesale and industrial businesspeople are used to working with larger sums of money. They do not freeze up when the salesperson starts throwing around numbers such as $300,000. Our prospects are right at home. Wholesale and industrial businesspersons operate their firms on a daily basis as we all do, but they think in terms of good and bad months, quarters, and years. Also, within this time frame, they can have seasonal peaks and valleys, as contractors do. Having to operate in this manner requires planning; therefore, our prospects will be more willing to listen to a good sales presentation and to appreciate ideas and help on future expansion.

Our prospects may have more employees than retailers, including, in a great many cases, delivery personnel. They will have all the problems that go with having people on a payroll and trying to get the goods delivered to the customer.

The contractor faces the same situations in trying to get the job done, and will have no trouble establishing a line of communication with this type of prospect, and through this, acceptance. Several years ago I was working with a good prospect who was in the wholesale food distribution business. I had already made two calls on this particular gentleman but felt that I was not really getting through to him. He seemed distant and noncommittal. I was having trouble obtaining the necessary information to proceed with working up a presentation. I hung in there because the man was a first-class prospect.

The bank next door to his location had bought the prospect's warehouse after many months of negotiating. He had been given 6 months to move. Two months had already passed when I came into the picture, so owner X was very quickly running out of time, a factor that made him a prime prospect. During the second meeting I had determined that activity eased up during the middle of the morning, so I had a third meeting scheduled for the following Monday at 10 a.m. When I checked into the office, it was a typical Monday morning, and one of our truck drivers had stayed out, creating a bad problem on a special job. I called, explained my problem, and pushed back my appointment. I finally solved the problem and showed up 45 minutes late. The first thing owner X asked when I stepped into his office was whether I had much trouble with drivers staying out on Monday mornings. It had been a hectic morning, and I must have been looking for a sympathetic ear to bend. I took 10 minutes and told him how it was really difficult to plan for Monday mornings when you didn't know who was going to show up.

After my tale of woe, owner X sent for coffee and proceeded to tell me about his morning. Two of his drivers failed to show. We spent the next half hour telling each other how hard it was to find good help, then the next hour working out the details of his new warehouse. He was a completely different person. Suddenly I wasn't an outsider pushing my way in, but a kindred spirit—one whom he could talk to and count on to understand. Two weeks later I had a signed contract for a 10,000 ft^2 warehouse, with a 2500 ft^2 attached office.

Dealing with the wholesale and industrial businessperson is much easier than dealing with the retailer. The former operates from an office behind a secretary. Most of the time you will be able to make an appointment and conduct your business over a desk in an efficient manner. This helps the salesperson because he or she will be able to put time to better use than standing around waiting for the prospect's customer to clear out. It is easier for the sales agent to drop by and hope to catch owner X in; and if X is indeed in, chances are that X will see him or her. Since the sales agent is not directly involved with the customer, the prospect has greater control over his time.

After making several contacts, the construction sales representative will come to understand and appreciate the wholesale and industrial prospect from still another point of view: the size of the job. These businesspeople need space to carry on their enterprise; the wholesaler needs space to stock goods, and the industrial person has to have space to manufacture wares. These people think in terms of 10,000 or 100,000 ft^2, and that's what makes the metal building salesperson sit up and take notice.

The important thing to keep in mind, when you establish a contact that you hope will result in a contract, is that the wholesale and industrial prospect is in many ways similar to the general contractor, and these similarities can be used. Better yet, they should be pursued and cultivated. The sooner a rapport is built, the easier the selling.

You may be thinking, why all the fuss over finding a common ground with the prospect? What's the big deal? Just talk football or golf, and do the same job. Try it, and you'll wonder what happened. The retailer has very little to nothing in common with a general contractor in a business way, so you maintain a pleasant, professional relationship and go from there. The wholesale and industrial people do have business areas that are very much like those of the general contractor, and the good construction sales agent will capitalize on them. But never try to make something up with sports or whatever. Realize that these people you are wanting to do work for aren't dummies. They are astute members of the business community, and they didn't get where they are without having some gray matter between the ears. They'll read right through a phony "hail fellow, well met" greeting, their off button will start glowing a bright red, and you will be out of the picture. Even worse, they might tell their friends about you.

Now there are exceptions. Remember, selling is an art, so there will always be exceptions. If the prospect brings up the subject, okay, pursue it—but not into the ground; don't keep flogging it. It's little points like this—knowing how far to carry a subject, for example—that make the difference between a good sales agent and a super one.

Professional

The professional category includes doctors, lawyers, certified public accountants, dentists, and so on. The general contractor has even less in common in a business way with this group than with the retailer. These members of the business community don't have hung-over drivers on Monday morning.

They seem to rely a great deal on other professionals when they need something done. Thus, when you are dealing with the professional, you cannot overstress the fact that design service professionals will be overseeing and designing the project. Don't hesitate to give out the names of your affiliates because your prospect may know them, in a business manner or socially.

These professionals are hard to see! It's almost impossible to get past the secretary on the telephone, and don't even waste your time stopping by if you don't have an appointment.

When selling, you have to play the percentages, and with respect to percentages the professional category doesn't offer as many business prospects as do the first two groups.

There are exceptions: professional people such as doctors, dentists, and lawyers are constantly looking for places to invest money, whatever the reason. Commercial real estate can be made attractive to them. The metal building sales agent can use this fact to put together a venture with the clear understanding that she or he is going to build the facility. If this segment of finance is new to you, then I suggest you make it a point to learn all you can because it will be a big help to selling.

To summarize briefly what we have said so far, the methods that the metal building salesperson can use to gain entry to the prospect are very important. It takes two to sell something, and you are only one, so you must be able to confront and hold on to the other person. We have discussed how to avoid turning off the prospect. The three general areas of the business community have been looked into, and their description will prove helpful in letting the salesperson prepare mentally for the particular type of prospect to be called on. Several more topics are discussed below. They are just as important to selling as anything that has been presented so far.

Don't Talk Down to or Put Down the Prospect

This admonition will probably be obvious to the reader, but it still needs to be brought out and should always be in the back of the salesperson's mind. The metal building sales agent calls on all kinds of people and has to be as adaptable as a foot soldier in the field. From the plush, well-appointed office of a trucking company president who wants to spend a lot of money to enlarge his building, the sales agent may call on the owner of a small auto repair garage. The office consists of one desk in the corner covered with paper of all kinds. It will most likely look as if someone emptied the trash can on top. All the paper will be smudged with grease, and right in the middle will be two or three wiping rags. The couple of chairs around the desk can be classified as early grease pit. You hope above all that you can do your business standing up!

Most garages these days will have a guard dog under the desk giving you a beady-eyed look that says, "You're here only because I let you stay."

Owner X comes up, and you introduce yourself and put out your hand. That's right, put the hand out; assume the owner's hand is squeaky clean. If the owner has dirty hands, most likely he or she will say so and let you know he or she would like to shake your hand except for the dirt. But if the owner proceeds to shake hands, then do it. When he or she points to a chair, sit in it; but as far as the dog is concerned, you are on your own—the best advice I can give is to ignore it.

Granted, this description may be a bit heavy-handed, but I feel it makes an important point. If you look around as if you can hardly wait to get away, or if you don't sit, or maybe if you pull out your handkerchief to wipe your hands—then bang goes the off button.

My experience has been that the self-made person is a little more sensitive to the put-down, no matter how slight. I'm sure a psychologist can go on for days about why, but I really don't have the space for that. It must suffice for now that you are aware of this situation. This selling area has to be played by ear.

Don't Overlook Selling to the Prospect's Employees

Whenever you get the chance, talk with the prospect's employees. It's surprising what you can learn that may be a big help in negotiating. For example, I was trying to sell a motorcycle dealer a new building recently. He planned to relocate and had a new site, so he was a prime prospect. I kept going back time after time to try to get the project going so that I could put together a proposal. Well, after two weeks of feeling pushed around, I was ready to drop owner X and move on to something else. I decided to make one last call. I phoned, set up the appointment, and showed up on time to find owner X at lunch. Business was slow, so I struck up a conversation with the sales manager. We talked about a number of topics; then I brought the conversation around to the new building. The sales manager was very excited about the idea of a new building. He had a head full of ways to increase the business. I asked for his thoughts on several showrooms and shop layouts that I was working on. The man was pleased as punch to help. Now, I honestly wanted his help, but at the same time I was selling myself to him. I wanted to find out if anything was happening that I should know about. I mentioned that the owner seemed unconcerned about moving along with the new building and was a hard person to do business with. Then the manager dropped the bomb. The building they were operating from had been condemned, the previous month, and they had to be out in 6 months. The owner's attitude was just her way of trying to get the bottom price.

My point is: I was ready to give up until I fished about and made a big catch, which was exactly what I intended when I started talking to the sales manager. So chat with the hired hands every so often, and you may get some useful surprises.

Be Conscious of Prospect Embarrassment

This area of discussion may be a little vague and may even cause some outright laughter. But there is such a thing as prospect embarrassment. I know because I lost a nice contract because of it.

When the sales agent is working with a prospect from the beginning, there usually are many changes necessitating a constant flow of information from the prospect to the salesperson. After asking the sales representative to change this and that for what seems like the tenth time, the prospect may begin to feel a little guilty about using the sales agent's time. The prospect who is not completely sure about going ahead may just back off and lose momentum rather than ask the sales agent for something else. The salesperson at this point may not be pursuing the prospect as actively as before, for some reason. The project just dries up for lack of communication; the prospect is too embarrassed to ask, and the sales agent does not follow up.

The next thing the sales agent hears about is owner X building a new place, and the agent wonders what happened. There is no one to blame but the agent, and chances are that he or she won't be aware of what really went on. All the fault lies with the sales representative for not making clear that the prospect should call on her or him no matter how many changes are necessary. Make the prospect understand that it goes with your job, you expect to do it, and she or he should not worry about your time. The good metal building sales agent should be able to tell if the prospect is serious or is just being manipulative, looking for information that can be used later. This is a judgment call, and with a little time and experience the construction salesperson will learn to make the right call.

The sales agent must keep impressing on the prospect the importance of always keeping the lines of communication open, no matter how much time is involved. The good sales representative will not depend on the owner to do all the work. The other mistake to avoid is failure to follow up. The prospect will not mind using your time so much if you offer it face to face. Remember, prospect follow-up is one of the most important aspects of closing a contract.

That job I lost was a warehouse for a heavy equipment dealer. It seemed everything was always being changed. I started on two other projects when I was working with the heavy equipment dealer, and I let things slip. Not hearing any more from the dealer, I thought he was still trying to decide what he was going to do. Two months later I ran into the dealer at lunch and asked him how things were coming along on his building. He looked slightly uncomfortable and told me he had bought from another firm. Before I could say anything, he went on, "I felt so bad about grinding up so much of your time. I didn't want to bother you."

I stood there dumbfounded! How could he bother me if there was a contract involved? Well, I chalked it up to having to deal with individuals and the fact that everyone is different. Then 6 months later I was having coffee with a friend who builds custom houses and performs a great many of the same services as a design-build metal building contractor.

We were comparing notes on selling construction, and he repeated a story that sounded very much like the one I have just described. He had experienced prospect embarrassment also. I have since talked with several other good construction sales representatives, and they all agree that there is such a thing.

If you are skeptical, that's all right. But when you are working with a prospect, tell her or him not to hesitate to call you. And follow up.

Stand Straight and Look the Prospect in the Eye

Do not feel that you are imposing on the prospect. Do not apologize for being there. Don't slip in the door and stand in the corner, twisting your hat, hoping finally some nice person will speak to you in friendly tones. Walk in as if you know exactly what you are doing. Look and be confident, and act that way. No businessperson respects a cream puff. And if you're wondering if you can do this and maintain the low-profile, soft-sell attitude, the answer is yes. It's not that easy to explain; it's something that just has to be done. I think the best advice is to practice the low-profile, soft sell, but give the impression that you are a businessperson conducting business during business hours and have no reason to be apologetic.

I like to take the direct approach sometimes when I know something about the prospect. For example, with the owner who has the reputation of chewing people up who call on him or her, after doing what has to be done to get to owner X and while shaking hands, I look him or her in the eye and say, "I'm here because I can save you quite a lot of money on your new building. It'll take me 5 minutes to tell you how." I have never had this direct approach fail. Owner X always takes the time to listen. What that direct approach does is reach out and grab the owner by the shirt. You have made the owner want to talk to you, and that's what selling is all about.

This direct approach will not be the best way all the time; usually it is best for the fast-moving fire-eater type. The trick is to figure out what type of person you are calling on, so find out all you can about a prospect before the initial meeting. Calling a mutual friend is an excellent way. If possible, observe while waiting; doing so should give you a good handle. Another nonapologetic approach is one I use when the prospect begins to talk money and feels maybe the price is too high. I explain the cost of construction, and then, making it a point to look the prospect in the eye, I say that *profit* is not a dirty word and that I fully intend to make a fair profit out of the job because that's the reason why I work—the same reason that the prospect needs and wants the new building: to make money!

I generally don't use this approach unless I get the feeling that the prospect is looking down at me; then I quickly yank him or her back to why we are talking and the fact that we are both after the same end—making a buck! Never apologize for wanting to make a profit. It's really the main reason you are there.

Remember, not only are you selling, but also you are operating in one of the truly specialized areas of selling. The metal building sales agent has to be a good salesperson and has to have a good working knowledge of the construction industry. She or he has to be able to understand zoning, sewer and water problems, drainage, the various trades and how they dovetail together, architectural aesthetics, site work, paving, construction finances, permanent financing, and so on. In short, the agent has to be thoroughly knowledgeable about a very large and complex industry. At the same time, he or she is selling in the big league. Most sales representatives close deals for $100 or $5000, but the construction sales agent closes contracts for $250,000 with the same ease that a tire sales agent sells $80 worth of tires. So hold your head up; few people can sell construction. It takes a special combination of knowledge, the hard hat in one hand and the calling card in the other.

I've only discussed the important areas of selling that apply to metal building sales. Other segments have their special dos and don'ts. For example, the factory representative calling on accounts on a regular schedule has certain rules to follow, since even after the new account is opened, she or he has to keep reselling every call. The real estate agent showing houses may be primarily selling to the woman of a couple, which also requires special knowledge. There are many other aspects of the selling profession as a whole, but we are only interested in the significant information that will make for successful construction marketing.

In closing this chapter, I want to point out that not everyone will be comfortable in this type of selling, and some will be comfortable only in certain parts; for example, the person who does not like cold calling may be first-rate in his or her own conference room. But please take the time to give it a good shot. Who knows—you may uncover hidden talents.

NOTES

NOTES

6
Maintenance and Repair Work

This area can be a very useful and profitable part of your metal building business. The place to start is to let the business community know you are the company to contact when there is a problem.

In general, the subjects in Chap. 2 will be addressed, but first let me define the two main areas to be discussed.

Maintenance and Repair Work

Maintenance and repairs that are of the everyday variety can be handled in the normal work schedule that works well with all involved. When the wall sheets are faded or the roof sheets have gone to big patches of rust and have to be replaced, the customer can easily understand the time schedule. Also while the customer is waiting, business is not seriously affected so the contractor's time schedule is not a big problem.

Now let me stress that getting the job done is not something to do when the work crews having nothing else to do. It's important to give the customer an idea of when you will get to work; after you have ordered the material and have a definite delivery date, call the customer and work out a time. Then show up and get the job done when you said you would. Nothing makes a contractor look as good as getting things done on time. Punctuality counts.

Also while the noted wall and roof sheet replacements can be a nice bigger job, a lot of maintenance and repairs will be small jobs that may take only a couple of hours or maybe a half-day or so. But the same time rule applies: Tell the customer when and then do it. When the referrals start coming in, you will know your time was well used. (See Fig. 18.)

Figure 18. Buildings needing maintenance repairs.

Emergency Repairs

Now let's look at the other maintenance and repair area: emergency repairs. Let me give an example. You have a business that uses a large metal building that you bought from the previous owner last year. You have no idea who the building contractor was, and since the building is

Figure 18. *(Continued)*

Figure 18. *(Continued)*

Figure 18. Buildings needing maintenance repairs. *(Continued)*

more than 10 years old, the contractor may not still be in business. Here is your problem: A truck backed into the sidewall of the warehouse and punched a large hole in the wall. Now since you don't know who built the building, you don't have anyone to call directly.

The first thing you do is call a friend and ask for help. You ask whom you can call. Your friend, who has some construction contacts, has no idea except the name of a firm that puts up metal buildings. You call the company and are told they are busy right now, but they may be able to work you in next week. Next week! You have a hole in the wall you can walk through, and you are looking at next week.

So you pick up the Yellow Pages, hoping you will be able to get someone to move faster. You go to the metal building section and read down the names, concentrating on the advertisements that accompany the names. These blocks tell how great their name brand is and how it can be used to do this and that.

Not one, though, says anything about providing a metal building maintenance and repair service, so you are at the same place now as you were when you reached for the Yellow Pages. The only thing to do is to start calling the names and see if someone can take care of your problem. Chances are, it will still be next week, so you had better get somebody in the warehouse to plug the hole with something and then back a truck up next to the damage.

Now suppose that when you looked at the Yellow Pages, you went right to an advertisement that listed all the things the metal building can be used for as well as emergency repairs and maintenance. You call, and the problem is taken care of. A crew will stop by this afternoon to cover the hole, and the final repair job will be completed when the matching wall material is delivered. You expect to pay a little more for the emergency close up, but it's money well spent, since you can now go back to running your business.

If you were the one called to do the repairs, you would have a nice job on the books, but the way to make this happen is to let the business community know whom to call.

Now let me give an example of a very serious emergency call that I answered; currently I am in the process of putting together a price for the insurance company. (See Fig. 19.) What happened here was that the steel loading platform that let the forklifts move material from the railcar into the warehouse was not removed before the railcar was moved. Thus the platform along the metal building slab was dragged for approximately 40 ft and proceeded to cut everything in its way. The first action taken was to support the two roof beams that had their columns pulled from the concrete, then close up the walls with plywood. Now the insurance company involved is taking prices to make the repairs to bring the building back to like it was, after the emergency repairs were done. My point here is that the warehouse people involved knew someone to call and get things moving.

As noted, we are going to use some of the subjects covered in Chapter 2.

Advertising

Yellow Pages. First, add to your Yellow Pages listing that you do maintenance and repair work, and emergency work if you want to cover this area.

Special Business Book. If your area has a business telephone book that lists all the different kinds of businesses in your area, then it will be useful to be listed there.

Direct Mail. This is good to use, so check it out. Maybe your metal building manufacturer has a place for maintenance and repairs in the direct-mail material. If not, consider coming up with your own direct mail.

Figure 19. Buildings needing emergency repairs.

Figure 19. *(Continued)*

Figure 19. *(Continued)*

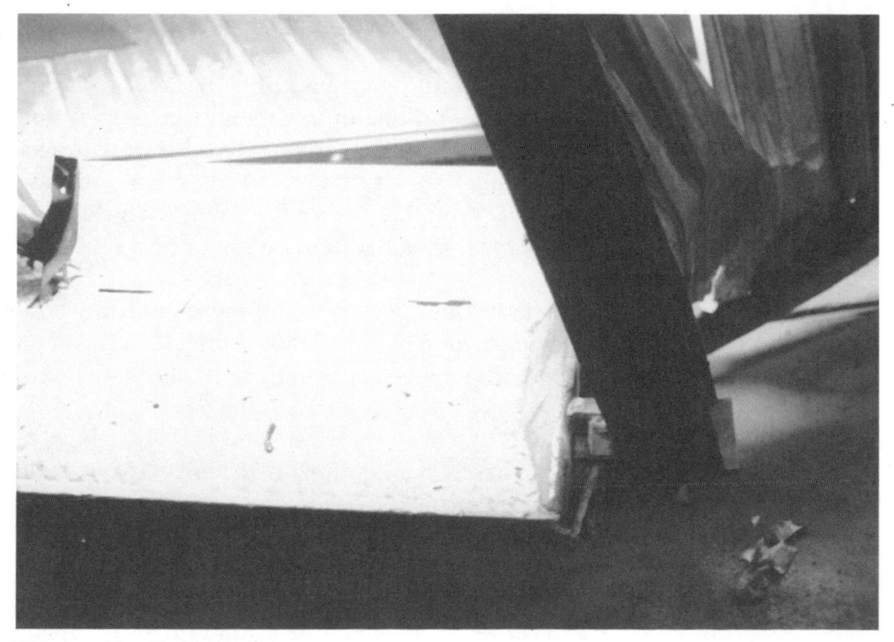

Figure 19. *(Continued)*

Figure 19. Buildings needing emergency repairs. *(Continued)*

Calling Cards, Stationery, Envelopes. I discussed calling cards in Chap. 2 and telling the prospect exactly what you offer. So let's go a step further and include maintenance and repairs on your envelopes and stationery as well as calling cards. Remember the boatyard owner I mentioned who showed me the maintenance problems in his metal building after looking at my calling card. Well, I am sure the owner would have preferred to contact somebody local if he had known about him or her.

Personal Contacts. Let's tie in this special area of metal building contracting with locating the prospect for a new metal building. When you let your personal contacts know what you do, bring up maintenance and repairs.

Referrals. This is a good resource, so make sure the owners for whom you have built know about this special area. And when you get a referral, make sure the owner who referred you looks good, by giving out your name.

Job Signs. People who have a metal building will take a look at one going up as they drive by and even may glance at the job sign. So list maintenance and repairs on the sign along with the other information.

Cold Calls. Let's go back to cold calls. As noted in Chap. 2, I picked a rainy, stormy day to make maintenance and repair cold calls on older metal buildings with good results, so never hesitate to make the maintenance and repair cold call. Often it will be combined with making a cold call to push a metal building, and once in a while the prospect who is already in a metal building will be more interested in maintenance and repair work than in adding onto the building.

Insurance Business

This is a special segment of the business community who will be very glad to know what you can offer: the insurance people. Let us return to the example of the truck knocking a hole in the wall. If the truck were owned by another firm, then its insurance would come into play, and the owner would address this at the time; and if the local insurance people knew what you did, the owner would be looked after. The insurance people will find you a very useful connection when they get a call concerning metal building repairs.

As a matter of fact, I was just involved in a similar insurance situation. A local florist who operates out of a metal building had a large delivery truck hit the corner of the metal mansard across the front of the building. This was not an emergency situation, simply a cosmetic one. I was called to work up an estimate to fix the damage, and hopefully, I'll be getting the go-ahead in the near future.

So tie into the insurance business for a good source of repair jobs.

The bottom line is to let the business community know what you can do. Then just don't sit around and wait for the phone to ring; go out and make something happen!

Subcontractor Network

I want to discuss a special area of maintenance and repair jobs that are not tied to the locating-business subjects we have been going over. The topic is working with your subcontractor network in two areas: overhead doors and heating, cooling, and ventilation.

The commercial overhead door company that installs the overhead doors in the warehouse and shop areas of a metal building is a very good source of repair work, and once in a while, it can be emergency repair work as well. Metal buildings use a lot of overhead doors. (See Fig. 20.) Trucks and equipment-moving machines are constantly moving about, so every now and then the door or metal building frame gets tapped. When this happens, almost always the door company is called to make things right. This is where it really pays to have the door company name and phone number on the door.

The door people get the call and show up, and the first thing they see is that no repairs can be made until the metal building damage is fixed. So they get the door down if it was up; or if it was already down, they secure it so the building cannot be entered through the door. Then you get the call from the door firm to check out the damage and get it fixed, so they can rehang the door. Now standing back is the owner, watching everything and tapping his or her foot while waiting for the people involved to get moving so the loading dock can be put back in service.

This is a situation in which you and the overhead door people have to work together to make each other look good, because if that owner is happy with the door repairs, the word will get out.

This network is a two-way street, and many times I am asked about the overhead door, be it maintenance repairs or adding a new door, and I

Figure 20. A damaged overhead door.

always give out my network overhead door company's name. Now I cannot recall how many times I have been brought in by that company to help with a door problem in a metal building; and even better, after I was brought in, I did some other metal building maintenance and repairs. I might add that most of this happened after I scouted out the building and brought the details to the owner's attention.

The heating, ventilating, and air conditioning (HVAC) subcontractor basically does nothing except punch holes in the metal building, and that's where you come into the picture. The real problem is equipment installation and roof penetrations. So I get called in to work with the HVAC people to make sure things are done correctly for the metal building. Again it's a two-way street, and when I am asked whom to contact for commercial HVAC work, I give them the name of the company I network with.

While overhead door and HVAC subcontractors are the main companies you will be working with, every once in a while all the other subcontractors will come up with a job for you to consider. So work out a two-way network with your subcontractors, and look after one another.

I have the perfect example of this arrangement. A week ago, the surveyor that I work with called and told me about a hardware store that was planning to expand, and it had bought the property next to its parking lot. While there doing the survey work, my friend asked how they planned to expand. He was told that they wanted a freestanding 2500 ft^2 metal building placed back in the corner so trucks could easily get to it.

The surveyor then told the hardware people about me and said he would have me contact them. He called me the same day, and I called the next morning and then went by and got the ball rolling. There is nothing positive yet, but I feel I have a very good chance to get this job.

Now, whether I do it or not, the important point here is the source of the prospect, my subcontractor network.

Special Work Situations

Now you have all this maintenance and repair work to do. Let's look at special work situations of your work crew.

The first thing to look at is timing, and how to work with the owner. We talked about the emergency call, but how do you handle this? It is simple. Let the emergency call take precedence over everything else. The most important thing here is communications. Let me give an example. I recently had an emergency call from a truck line for which I built a warehouse. Again this concerned a 14-ft-wide by 16-ft-high overhead door. I called the steel crew on another job and told them to get over to the truck line and take care of things. I then called the owner of the building they were working on, which was an addition to the existing warehouse, and he was literally watching everything that was being done. Anyway, I got the owner on the phone and explained the emergency situation. His reaction was what I have experienced a number of times. There was no problem leaving his job for the reason the crew left. The owner liked how emergency problems were taken care of because he could easily be on the other side himself one day. So do it, but let everyone know what is going on.

Another timing problem is tied in with the business process where you are doing a special kind of job. Let's say your people will be working on the roof, and you don't want anyone under the area while the work is going on. Now the business using the building manufactures items on a production line, and having no one underneath the roof work means shutting down the production line. That's the last thing the owner wants to do. So you have to do the work over the weekend when everything inside is shut down.

The best way to handle the work time is to stay in charge. While you are talking with the owner about the work to be done, take note of what the building is being used for, and then you bring up the subject of special work timing. You will always get a nice reaction from the owner and a lot of smiles from employees if they are part of the conversation.

Every now and then, there can be something strange that presents a problem at the work location, and the owner cannot take care of it, and you have to handle the situation. I have two good examples.

The first occurred at a manufacturing company for which I had built two metal buildings; I also did maintenance and repair work on the other large metal building the company started with. The job consisted of making some repairs where a wall and roof came together. Now here was the problem. This company made items that the government used in the space program, and the cold war was on, so security on some of this stuff was unbelievable. Since the work crew had to be inside as well as outside one of these high-security areas, the job was scheduled to be done when the room was not being used; and even then there was a security guard in the room, watching the crew. It was another weekend job.

The second example combines several things we have gone over. The business this time was a very large storage warehouse and truck line that receives, ships out, and delivers shipping containers from ships coming and going overseas. This company had more overhead doors than you would believe, and my overhead door subcontractor did all the door work. So I get a call that we had some door work to do. The shipping company wanted to make four doors that are all in line at one of the loading sections 2 ft wider.

So work had to be scheduled by the overhead door people and me because each door had to be completed in one day so that the opening could be locked at night. What the overhead door company and I did was to combine our work crews. My steel workers helped take down and put back the new doors, and the door crew helped when the steel crew enlarged the frame and set new anchor bolts. The owners of the shipping company were very impressed.

Now I'll tell you the problem, and it has a funny ending. Because this shipping firm both shipped out and received incoming containers from ships, it had its own customs office and warehouse, which was off limits to outsiders such as the door and steel crews. Now since no one could go through the customs area, the doors had to be reached by "going around your elbow to get to your thumb" and using their forklifts. We could not take our trucks in the building back to the doors because the drive-throughs would take only forklifts. And before you ask, I'll tell you why we did not park the trucks outside the doors. The doors serviced railcars with no road access at all.

Now here's the funny part. The roundabout route the work crews had to follow took them through a section of the warehouse where there were probably a million beer bottles from a well-known U.S. beer company stacked on pallets about 6 ft high. My curiosity got me. Where was all this beer going, I wondered. Well, it was obvious the cold war was over because this beer was going to Russia, I was told.

So both the door crew and the steel crew went back and forth through a warehouse full of beer. I still hear stories about this, and I imagine the friends of the work crews have heard some interesting stories, too. Plus, I would like to think none of the beer was tasted, but frankly, who knows?

Now these examples point out one thing to always keep in the back of your mind. Maintenance and repair work takes your crew into buildings being used, so you will have to coordinate when the work is done. Also in this case, it helps your reputation if you give in a little to keep the owner happy.

Often you will not have any timing problems, but when one arises, don't let it turn you off. Use it to make your company look good, because the payback will start coming in, and that's the bottom line.

Work Crew

There is another important subject to address concerning maintenance and repair work, and this area also applies when you are adding onto an existing building.

Construction work crews and especially commercial crews are not known for their well-spoken, polite use of words. Most of the time it can be a bit rough, and when other people are around, this can create a problem.

I have a good example. After I had been in business a couple of years, I was asked to put an addition on the side of a building I had built. It's an auto parts business, and they wanted the showroom to be enlarged as well as the warehouse area. So after the metal building was erected, the joining wall was removed at the showroom. While the showroom area was being finished and tied in, there was a steady flow of people buying car parts, and we had a lot of lookers.

I was in the old showroom, talking to the owner whose son was waiting on a customer, and his wife was in the office working at the desk when from the work area next door someone shouted really bad language, heard by all on my side. I have never forgotten the look on the owner's face when this happened, and I was truly embarrassed beyond words. I ran next door, put things in order, and came back and apologized to everyone who heard. The work crew looked bad, but the situation also made me look bad, and I did not like that one bit.

Since that embarrassing incident took place, I have made it a point to make sure crews working around outsiders do not use bad language.

Decent language is not the only area to take note of when you are working in the middle of the business. Simple politeness when dealing with the employees, customers, and especially the person in charge makes your company look good.

Plus there is another look-good area. Check the appearance of the work crew. Now I know construction work clothes are supposed to look used, but when your crew is elbow to elbow with employees or customers inside the building, there is no reason for them to look as if they just poured some concrete; or when they brush up against something, leave a smear of dirt, grease, or paint. Add to this area the use of restrooms. Often when doing a small job or fast job, your crew can use the company rest rooms. Again, make sure the company reputation remains good.

One last point while we are on this subject. When I speak of crew, I mean yours and your subcontractor's. This is always the area where you are in charge, so do it.

Now after all these negative points, I have a good point to consider if you have not already done so—company uniforms. When your crew has a company hat, jacket, sweatshirt, or T-shirt with the company name easy to read, they start out looking good as soon as they walk in carrying tools and materials.

Maintenance and repair work can be a profitable part of metal building contracting. This chapter has covered getting the work, handling job situations, and making you and your company look good. The remaining chapters will also cover maintenance and repair as the subjects apply.

NOTES

NOTES

7
Design-Build Contracting for Metal Building Work

The design-build concept works very well for marketing the metal building. This approach can offer greater profit per job coupled with fewer problems during the building stage, and it will create repeat work for the future. Sound too good to be true? Is there a catch? Yes. You're going to have to look at the construction business in a new light. While hard work is part of the picture, the really vital area that you will have to understand is what must be mastered mentally. Working hard is very important, but working intelligently is absolutely crucial. To that end, this chapter provides some basic education.

Definition

Design-build construction is a partnership among the construction contractors, design professionals, environmental consultants, subcontractors, and materials suppliers. Through this partnership, the design-build contractor has the capability of offering a complete construction agenda to the owner, which translates to positive benefits for both the design-build contractor and the owner.

The contractor will benefit from stronger relations with partners, which will enable the contractor to provide quality service in a more timely fashion. The owner will be pleased with the hassle-free, maximum-value construction provided by the design-build team and will become a good source of future business for the design-build contractor.

Once the design-build team has been secured and the job won, the owner becomes part of the team. The owner can then utilize the contractor as a single source of responsibility for the job. The owner's business concerns will be addressed more quickly and effectively, making the entire process go much more smoothly. This will then translate to more efficient use of the owner's investment.

Sales Territory

Design-build business is not something that you can do in a hit-or-miss fashion. It is not to be pursued when things are a little slow in the bid market. To be really successful, design-build should become a full-time part of your business, and then it may become the main area from which your contracts come—and in some cases, the only area. I'm speaking from personal experience. I do no formal bid work; all my business is generated from the design-build concept.

The overall strategy or goal for the design-build contractor is to create a *sales territory* in which a large number of customers depend on the contractor for certain goods and services. A construction contract is never viewed as a one-shot deal, but rather as one of a line of ongoing contracts for the customer. Granted, it can be years between building contracts, but you should treat the customer as if another contract were coming next month. You should have enough customers in your territory so that something is always coming up.

Add to this the maintenance and repair work. While it may be years before you build a new building for a previous owner, the maintenance and repair work will be an ongoing source of work.

I have three good examples. I have had five contracts with a local seafood processor (three buildings and two interior jobs), three contracts with a metal fabrication shop, and three contracts with an auto repair business. Note that three customers gave me eleven contracts consisting of eight metal buildings and three interior jobs. Most important, in each instance, after the first job had been sold, all the repeat work was contracted with me as the only contractor in the picture. Also in each case, no one (and especially my competition) knew anything of these jobs until after I had picked up the permits; also there was ongoing maintenance and repair work.

In my opinion, repeat work is the best result of design-build contracts. It's sort of like a bonus. While a second job for an owner is always nice, I really enjoy it when an owner recommends me to someone else.

Now the idea of a sales territory is very much a part of selling metal buildings. Your building supplier will define your sales area, so it works out well to use the same area to push your construction business.

Design-Build Capability

To sell an owner on the design-build concept, you will have to have design-build capability. Don't think for a minute that you will be expected to put on your payroll the needed people to have design-build capability. The size of the construction company has nothing to do with design-build capability. Granted that it's convenient having the needed people in the same office, but the one-person operation with a secretary can have the same capability as the large office full of experts.

Design-build capability will require new business attitudes, in-house or outside professional services, and the creation of a database of subcontractors you can draw upon to help pull together the data necessary to sell the job. In other words, you'll have to make a construction team where you're the captain as well as the coach, and it's your responsibility for this team to be a winner.

Everything from here on has only one purpose: to help you become a winner in the design-build section of the construction industry.

Design-Build Team

At this point you have a prospect interested in the design-build concept, and it is now up to you to provide the necessary information to proceed with selling the project. You must use your design-build team to pull together the numbers.

First, someone (be it you or an estimator) has to have a very detailed understanding of what goes on in a construction project. I'm making the assumption that you, the owner, will be the team captain.

We've now nailed down the most important person on the design-build team, and along with this comes the responsibility of putting all the parts together to make the total. This team will have architects and engineers. Your relationship with the design people is very important. The design-build contractor is selling professional services as well as construction work, so it's vital that these professionals be comfortable with the team arrangement. These are the people who will protect the public and will determine exactly what must be done to adhere to all building codes and regulations. This fact cannot be stressed enough to the prospect during the selling stage.

In my area, certain commercial buildings that are not large do not require an architect's or engineer's stamp. However, I strongly stress to my prospect that my plans will be provided and stamped by a professional, no matter the size of the building. Also, if I know my prospect is talking to a competitor, I push the point that this competitor should make it clear exactly who is providing the plans. This is a good selling tip to show your prospects that in dealing with you, they are in effect dealing with professional design people as well. So an important part of your team will be the design people; and as a contractor, you have enough contacts in the field to work out some arrangements.

The other part of your team will be people directly involved in the building trade areas, that is, the subcontractors and materials suppliers. These people will provide special help for a particular area of expertise.

As the pricing process gets beyond a simple square-foot budget, which is where the numbers start for the prospect, particular information and pricing have to be provided. For example, there could be a site problem, and the site work people can be brought in to help.

Thus, the design-build team is made up of a team captain, the design professionals, and the building trade people. When necessary, other experts may be included, for example, environmental or hazard waste specialists. Keep in mind that your team will never be fixed, but always in some state of flux depending on the requirements of a particular project. It will be up to you as the team captain to put together the team and make it work smoothly and effectively.

Environmental Construction

Environmental concerns are starting to impact the business community very strongly. I am sure that these concerns and requirements will become even more demanding in the future. Therefore, I think environmental construction will offer a very profitable area for the design-build contractor to pursue. The reason for this is the extremely difficult road that must be followed to obtain the necessary permits to construct anything that has an environmental problem.

What the design-build contractor can offer the business owner is single-source responsibility in an area that can be completely overwhelming during the plans, permits, and building stage of the project. I firmly believe this area has limitless possibilities, but the design-build team has to be expanded to include the environmental professionals in order to take advantage of the opportunity.

Let's go through an example. You are talking with the owner of a manufacturing firm that uses special chemicals and stores them in 55-gal metal drums on the ground behind the building. The environmental protection people have checked out the situation and told the owner to get the drums off the ground and into an environmentally safe storage building. The owner knows he or she had better comply.

Now I digress for a moment. How you happen to be talking to the owner is also important. Did the owner find you in the Yellow Pages, or get your name from a friend? Maybe you made a cold call to offer this special service. Anyway, chances are very good that you are there because of what you have done to promote business. In other words, you made it happen, and that's very important.

I go back to solving the problems. You determine exactly what kind of chemical compound you are dealing with and how much has to be stored at any one time. Then you note exactly how the drums are handled—whether they are rolled around by hand, maybe a forklift is used or a manual hand trolley, whether they will be stacked or racked, and so on—until you understand exactly what will be taking place in the proposed storage building. Next you work out whether the building is at ground level or truck dock height, and decide exactly what the dimensions are. Once you have all the data, you tell the owner to give you a few days to pull everything together.

Your next undertaking is to meet with the environmental professionals on your design-build team and to pass on all the information. Let them do a little homework and provide you with environmental information that affects construction. For example, the chemical compound is explosive; therefore, explosion-proof lighting and blow-out wall panels may have to be used. But the real problem is leakage, and all precautions must be made to contain any leakage so it cannot get into the water table. This means curbing placed in the concrete floor with a clean-out depression so that leakage can be collected and cleaned up. Also, the concrete will have to be sealed.

With this information in hand, you check with your team architect and merge the construction requirements with the environmental ones. You can then work up a budget to get things going. But be forewarned about the price—it can really skyrocket. The example I have used is one I recently priced, and the cost was three times that of a normal storage building of the same size.

As this example shows, the absolutely vital information is provided by the environmental professionals. Also they may have to be brought in during construction to check or even certify a particular area. These people can become a very important part of your design-build team. They can also become the source of a lead that can turn into a nice job.

The environmental study is done before anything else, even before the building site is purchased. So the environmental people can then recommend a firm with a track record in environmental construction that will make things move faster and more smoothly for the owner. I strongly suggest you give this special area of construction some thought and make the environmental professional a part of your design-build team.

Project Steps

Your firm is now in the design-build business; you have convinced the prospect that you have design-build capability, and it's time to start putting together the numbers to see if this project can be signed up. There are four basic guidelines to follow.

The first step is to grasp exactly what the project is: building size, free-standing or an addition, and so on.

The second step is to factor in the items that pertain to this particular project. For example, are there any foundation problems to consider? How about the architectural treatment and mundane things such as overhead doors and loading docks? Then there will be office layouts with maybe a private bath for the owner. All the information you pulled out of the prospect is used here. You add it all together, make some preliminary drawings (nothing fancy is needed here, just simple line drawings will do nicely), and start plugging in the numbers. I find square-foot cost to be very handy. At this point you begin to get a handle on the dollars and have a good idea of how you stand with the budget, if there is one.

These first and second steps can run the entire gamut of construction from a simple box to a high-technology laboratory, but from my experience most are fairly easy to pull together. Now here's where a good database of construction costs will really pay off. You'll be able to pull together the needed information in a minimum of time.

The third step is to bring in special help for a particular area of expertise. For example, the prospect wants a special overhang to protect customers as they use the front door. So you fax a sketch to the team architect for comments. I am not talking about a lot of time, simply a few minutes invested in the sketch by the architect, and then you get the information.

You may have to touch base with the engineers and then the site work people. Next there are the electrical, heating, and air-conditioning people, and the plumber must always be brought into the picture. You may feel your database can cover all the sections for you at this point, and in a few cases you'll be correct; but from my experience, I've found every job to be a little different. To cover myself, I don't ever hesitate to bring in help.

In the third step the design-build concept starts to get a little tricky. In steps 1 and 2, you and your employees were the only people involved, and you had complete say over what happened. Now you're going to bring in outsiders and get them to donate their time to you for a project that may not be built.

During step 3 the subcontractors are important because you are working from sketches and line drawings, not finished plans from the design part of the team. That will come later, after the job is signed up. For now you are working on a budget estimate to get the selling process moving, which is where the subcontractor comes in, and you may want to bring in a supplier to furnish information and prices. This will be especially true when you are pricing the metal building.

You will be mostly using the subcontractors at this point, and you must have complete confidence in them. When the electrician tells you, based on the building use and shop equipment loads, it's going to take an 800-A panel and it's going to cost $X, it's important for you to feel comfortable with this information.

Let me emphasize that right here is where the design-build concept becomes a two-way street for every member of your team. The people you want to work with will have to know that if they work with you at no cost and provide information, they will do the work if the job sells. And that, frankly, is all there is to it.

Your job as team captain is to pick the subcontractors that inspire your confidence and then sit down and work out what you all can expect from the association. It may be entirely new to you to sit down and discuss business with a subcontractor. You may think that to deal with subcontractors requires you to yell questions in the phone while you reach for your antacid pills with your free hand. Bid jobs seem to bring out this scenario; but, as I stated, design-build means fewer problems during construction, which means better relations for the general contractor with subcontractors.

There is also a positive side effect for the subcontractor, especially in the electrical, HVAC, and plumbing fields, that can come from a design-build job. This is the service account. Design-build jobs are people jobs. That is, the owner and the contractor have a close working relationship, and the major subcontractors also become part of the picture.

For example, the owner will bring up an electrical change that her or his shop foreperson has suggested, so everyone meets, and in short order the electrician is on stage. If the owner likes what you have provided, chances are very good that when service work is needed, this electrical firm will be called in. I've seen this happen many times. If the subcontractor in question does not pick up on this, then point out that it's to her or his advantage to make that owner glad the subcontractor is on the job. This makes three things happen. You look good by having first-class subcontractors, the owner is happy about the quality of people on the job, and the subcontractor is creating a potential service account. Everybody wins!

Added to these three positives is the fourth that evolves when the subcontractor picks up a service account. You get a source of information on the inside who keeps you up to date on future work. All it takes is a phone call from your electrician, telling you he or she was just in so-and-so's to look at moving some wiring for a new piece of equipment that may require more shop space. You may have it already covered because the owner was very satisfied with your work; but nothing is guaranteed, and you can never have too much information about future work.

You set out to create a business association of trades that you can trust and depend on, and you want to feel comfortable with the information they provide. I stress the team concept. Often after I have met with the electrical and HVAC subcontractors, they work together to define the areas where they interface; and it's done in their office, not mine. It will be an advantage if you can take personality into account when looking over your list of subcontractors, because this association will have to be made up of people who can work together.

To quickly reiterate, step 1 is to pull together a general overview of the proposed project. Step 2 is to factor into your overview all the special requirements for the job. Now frankly, steps 1 and 2 go hand in hand, and in many cases they sort of merge. Steps 1 and 2 generate the general specifications and a preliminary line drawing that in turn will be used to bring the needed experts into the picture. Step 3 is to bring in team members to provide budget estimates. When you have all the outside figures, you can then proceed to the final step.

The fourth step is to establish a price and present your proposal to the prospect. This proposal, be it a beautifully bound pamphlet or a page from a yellow legal pad, is the starting point for the project. To finally end up with what the prospect wants may require redoing the specifications and drawings. Make sure everyone understands that the first pass at the job will probably not be the only one.

Don't be discouraged at this point. I know it seems as if days and days will be necessary for you to work up this budget estimate. But once you have done a couple, you'll be pleased to discover the time involved is hours, not days. With all the marvelous gadgets we have today, you can carry on your business from anyplace. For example, you can fax the specifications and line drawings to your subcontractors and they can return their figures in the same way. Or, if you're sitting on the job site with some time on your hands, pick up your car phone, open your briefcase, and go to work.

Remember, you have to make something happen. The design-build team is only as good as its captain. Somebody has to be responsible for the end result, and that somebody is you.

Security

The construction sales representative should concentrate on job security and should not worry about what anyone thinks about being security-conscious. There are a number of ways to work the job without giving out any information about the identity of the prospect. Preliminary drawings are usually a dead giveaway. It's only natural to cover the sheet with the prospect's name, if for no other reason than to impress owner X. Don't do it; fight the impulse, and leave it off. Also, while we are on drawings (and I'm talking about preliminary drawings or sketches that you will be using to get budgets from subcontractors, not architect-prepared plans; they come after the prospect is signed up, and then security is not a problem), don't spell out what the building will be used for. For instance, if you are working on a retail auto parts store, don't spell it out. List the job only as "retail store." That's hard to pin down; "auto parts store" can be run down by a smart construction sales agent in a matter of hours.

You may ask, How does the information leak from the subcontractor? You have a good relationship with the subcontractor, who is not going to talk. But chances are, he or she will have to contract a supplier; there will be salespeople calling on her or him so he or she can put together a price for you. Your subcontractor may be able to keep silent, but office security may be sloppy. All an outsider has to do is to look around the desk or drawing table and spot your drawing; right then, that person is in the know, and your prospect is on the street. Contractors' offices, those of both general and subcontractors, generate a lot of activity. There are people coming and going all the time, and anything left out becomes general knowledge very rapidly.

Security was a one-person problem when you were working alone; when you have to bring in an outsider, security is still up to you. So, don't let your drawings give you away. They are one of the best ways of revealing information, since it is recorded on paper to be read by all. A few simple precautions will go a long way in this regard; use your head and don't be careless.

Also, when you are talking to anyone, including subcontractors, refer to the job in general terms. The prospect simply becomes *the owner* when I'm discussing something that pertains to the job.

General contractors are extremely lax in internal security. Internal security is very important when you're doing design-build selling. The best rule to follow when you're trying to sign up a prospect is right out of the spy handbook—what they don't know they can't tell. Although I'm sure you don't have to worry about industrial espionage, what concerns me is the occasion when the steel foreperson is having a beer with her or his opposite number who works for your competitor. Shop talk will almost invariably ensue, and all your employee has to do is to say that the company is working like crazy to sell a new warehouse to a moving and transfer company. It would be said in all innocence, but there's an excellent chance that word would get back to your competitor. I know this sort of thing goes on. The people in a certain trade all know one another. They seem to float around, working first for this company and then for that one. Often these people will work together for a while, then for competitive outfits, depending on the workload. But they still know and see one another as well as discuss their work.

Make sure your prospect is not the main topic of discussion at the local tavern next week. Any employee can unknowingly blow the whistle on you. The secretary going to the post office or the bookkeeper making a deposit at the bank can pass the time with someone from another construction company. I feel the real problem is that construction company employees are a little more concerned about job security than are people in other jobs; therefore, the uppermost topic is, How's your workload? They're not prying, just interested. If one firm's contract load is off, they want to be reassured by hearing someone else's is also off, which probably means the industry is off a little. Misery really does like company. On the other hand, if they are loaded with work, they want to brag. The net result is that the workload is talked about often, and in such conversations the cat can be let out of the bag.

It's impossible to keep your employees in the dark, but you can impress on them that their jobs may depend on their keeping their mouths shut concerning prospects and upcoming jobs. Maybe it wouldn't hurt to suggest they listen a little more closely to other people and bring in a lead, so that they might be able to pick up a nice bonus if a sale is made.

I feel the best way to handle the internal situation is to explain to employees that what they hear in the office is confidential. Next, work on a need-to-know basis. If you have an estimator to work up the prices, make sure she or he doesn't run through the office telling everyone. At this stage of the selling procedure, the estimator is probably the only one who needs to know anything specific. The rest of the office needs to know only that the estimator is working on a building price. As the selling process moves ahead, other people will be brought in. The bookkeeper will have to charge the estimator's time to something, so this person gets in the picture. Finally, the proposal has to be typed, and suddenly the typist knows almost as much about the job as you do yourself. Hopefully owner X signs the contract soon after this.

Along with the office help, you'll have the chief field people stopping by the office, and it's only natural for them to have a cup of coffee and chat—with you, or the estimator, or other employees. You may stop by a job and find that the conversation turns to work in the pipeline because the foreperson is vitally interested in the future workload. This will often happen if you're working as a construction sales agent for someone else. Anyway, it's hard not to let the key field people in on what you're doing. Tell them anything you like as long as you don't compromise the prospect. This takes some diplomacy because you don't want the foreperson to get the idea that you're hesitant about talking. I can't tell you what you should do. Only you know the foreperson in question well enough to handle the situation.

There's a natural tendency to want to give your people a lift by telling them what is going on when the future looks bright; but don't tell all if it might come back to haunt you. Learn to think about your external and internal security. Good prospects are not that easy to come by, so give yourself every break to sell the job. The same sales manager I spoke about a few pages earlier told me something else that I'd like to pass along. He said, "You never learn a thing talking." And I'd like to add my own rule: You'll never give anything away when you're listening.

For your own sake, prospect security is extremely important, so practice it.

Remember, you are the team captain for the design-build team, so take over and run it so that all the team members make money along with you.

NOTES

NOTES

8
Special Inspections

There is a new area that the municipal building code officials are bringing into commercial building construction that must be understood and considered. Special inspections is the area I am talking about, and this chapter covers the details.

The special inspectors are qualified engineers that act as a third party to oversee the structural side of the construction project. The special inspections are an added cost to the project regardless of whether you, the contractor, or the owner directly pays the cost; but at the same time there are advantages that are worth the cost. These will be covered later in this chapter.

The metal building field is very open to special inspections. First the building plans from the building manufacturer have to be checked, then the erection of the building is checked as it goes up, and the concrete it sits on is checked as well. So special inspections are an ongoing part of metal building construction.

Now let me put the special inspections in their place in relation to all the other building regulations. Say you are going to build an 8000 ft^2 storage building for a wholesale company. The facts are passed on to your architect who brings in any needed engineering help, and at the same time technical data from your metal building manufacturer are supplied to the architect, so everything is put together from the foundation up.

When the plans are finished, they are stamped and given to you. Along with the architect's drawing, your building manufacturer will supply you with a stamped drawing detailing the metal structure.

With these stamped drawings, you start the building permit process by turning over the plans to the code officials. Stamped drawings for a commercial building will go through the paperwork with no problems concerning the structural aspects, because experts have put themselves on the line.

So the permit process may drag a bit, but not from a structural viewpoint. The stamped drawings are what the code people want to see. You get the building permit, and you are a registered, licensed contractor qualified to construct the building. And while the construction is underway, you contact the code people to inspect certain areas to make sure the construction is following the plans. And the reason all this is taking place is to ensure that a safe building is being constructed.

Special Inspections for Construction

Now add to this process the special inspections that oversee everything I have just described. Everyone involved is checked out, including the code inspector.

I have a good example of what I am talking about. I have just completed a 50-ft by 100-ft, 18-ft-high metal building with 1200 ft^2 of office space in the front. After I started the building permit process, I contacted a special inspection firm that I brought into the picture. I gave the firm complete architectural and metal building drawings to be checked, and I gave them my construction schedule so the inspections could be put into the network.

The first thing checked was the soil condition under the building. A geologist punched down 15 ft at each corner of the building, and the soil was okay to build on. Then, while the foundation and slab form work were being done, the bottoms of the perimeter foundations and the condition of the compacted sand under the slab were checked out. The foundation bottoms checked out fine; but the slab sand did not pass, so it had to be compacted once more, and then it passed.

The next area to address is the placing of the reinforcing rebar in the foundation perimeter with special mats from the rebar positioned in the anchor bolt locations.

The first thing checked on the rebar is written material from the supplier plus the tags on the bar itself stating the rebar meets certain structural standards. Also the inspector and I kept some of the tags for future use if necessary. The next step was to place the rebar in the correct positions before the concrete was poured and then to have the placed rebar inspected and okayed, before the concrete was to be poured. At this time also the city code inspector checks to see if all is correct, and if so, gives permission to pour the concrete.

So let's get the foundation and slab poured. For this job, it was poured all at the same time since the job was considered a monolithic concrete job. Now while the concrete was being placed, special inspectors were on the job, first taking tests to see whether the concrete coming out of the trucks met the called-for strength, and while this was going on, cylinders of concrete were taken to be tested in 28 days to make sure everything met the required specifications, which by the way it did.

When you have a good-size concrete pour, the special inspection people will be there along with everybody else, so it makes for a busy day on the job site. Let me say that not all concrete pours are inspected, so check with both the code people and the special inspectors to know when to have the special inspections. It's mainly size—2 or 3 yd, no inspections, but for 50 yd that will support a structure, special inspectors are needed. So do a little homework, and check out the requirements.

The next area in which the metal building contractor has special inspections is the erection of the metal building. All structural aspects are checked, even the number of screws used in wall sheets. Also the many nuts attached to bolts are checked to make sure they are tight enough. Plus other structural subjects are checked out during the erection process so when the special engineers sign off that all is okay, everyone is comfortable that the building will keep standing.

Contractor Benefits from the Special Inspections

I am sure your first reaction to what I have just described is simply that this is another obstacle the contractor has to handle while doing the construction work. But the good side far outweighs this negative reaction. The special inspections are a helpful benefit to the commercial contractor. They are another area that backs up the contractor and offers another protection zone to run the project through.

The architectural plans and metal building plans from the building manufacturer are stamped and properly certified. Then the building permit people approve the drawings and give you the building permit, with the knowledge that the construction process will be inspected by the code inspectors to make sure the plans are being followed. So the contractor is required to build the building as the experts show on their plans, with code people looking over the process to make sure the plans are being followed.

The bottom line, however, is that the contractor is dealing with the owner of the building, and that means the check-writing owner deals with the contractor directly on all construction items including the problems. No matter what problem arises, the contractor is considered responsible.

This is where the good points of special inspections come in. The inspections provide another layer of experts who will check out the structural aspects of the building, and this makes the people who supply structural materials and do erection work more safety-conscious.

For example, the concrete supplier knows that the specifications call for a certain load-bearing capacity, and the special inspections will be done to make sure the specifications are met.

Then while the steel erectors are putting up the building, they will be checked out during various aspects of the erection process, and the special inspectors will sign off on the whole job when finished.

Now what this gives the metal building contractor is another protection zone so that a serious structural problem will not pop up after the building is constructed. I am not talking about the nickel-and-dime problems all construction jobs seem to have, but serious big-dollar problems that show up after the building is constructed. So while the special inspections are another area that the contractor has to contend with, they are well worth the effort.

Let's look at the special inspections from the design-build business side of contracting. While I am in the selling process with the owner, I bring up the special inspections, and I tell the owner there will be an added expense to the job; but then I explain all the good things about the inspections, and they are another backup checkout zone for the owner's protection. Since the owner is building this metal building for the long-run money-making process, the idea of an added layer of protection by certified professionals is well received.

This makes me and the owner happy because I have justified the extra expense, and the owner is more comfortable with the building program. Quite often the owner will stop by when impressive construction such as pouring concrete or erecting steel is taking place; and if the special inspectors are there, the owner is made aware, and it really looks good when you and the owner look over their shoulder. So take advantage of the special inspections and work them in with the design-build method.

Now let's look at the special inspections from another viewpoint. What I have described in detail so far was brought into the picture because the code officials added it to their requirements. In other words, you have special inspections performed because it's part of the construction process and you have no choice in the matter.

But let's look at special inspections from another direction. Suppose your work area does not have these requirements, so you are not required to bring them into the picture. However, you are still responsible to the owner for a structurally sound building, so you may want to consider bringing in special inspections to make sure you have another protection zone, especially when the job is large and the inspection cost can be easily added to the construction cost.

My point here is that even when special inspections are not required, you may want to consider adding them to the construction process, so make the special inspections a part of your design-build team.

The next area to bring into the picture is the codes themselves. I want to thank the *Building Officials and Code Administrators International, Inc.*, better known as BOCA, for allowing the special inspections chapter to be included in this book.

Chapter 17, "Structural Tests and Inspections," from the BOCA National Building Code, is located in App. B, page 227. If you are not familiar with what this chapter covers, then I suggest you look it over carefully. Take in the details, and don't be surprised when you find yourself going back many times to check something out.

To sum up special inspections, they are a growing area in the code requirements as more and more areas are adding them to the requirements. So if your area does not require them now, there is a good chance they will be added in the future. So you, as the contractor, should know the details and how to use them.

One last subject needs to be addressed before we end this chapter, and that is how your subcontractors and suppliers fit into the special inspections. Now we have already brought up the subcontractor network, but this is a very special area.

Don't take the approach with the subcontractor or materials supplier that the special inspections will be used to protect you, the general contractor, from their messing up and creating a big problem. Take the approach that the special inspections are a benefit, an added protection zone for you and the subcontractor and supplier. The inspections do not protect you from them; the inspections protect you and them from the structural problems.

In other words, you work with your subcontractors and suppliers as members on the same team, and this approach will keep the personal connection which makes things run more smoothly.

So the special inspections are really a benefit to the metal building contractor, and this approach will fit in nicely with the design-build system.

NOTES

NOTES

9
Selling the Building

As defined in Chapter 4, the prospect is convinced it would be good to work with you on a building project. However, while the prospect feels comfortable with you, she or he knows next to nothing about your company.

So your next undertaking is to sell the prospect on your company, and you do this by showing owner X all the facts and information that proves your firm has done work in the area where the prospect's building will be.

Selling Your Company

You've now sold the prospect on yourself; the next step is to sell her or him on your company. In other words, you have to prove to the prospect that you can do what you say you can do. All owner X has from you is a lot of words, and your words have not cost a cent so far. But can you perform? The best way to answer this question is to show the prospect past jobs and describe some of the details as well as for whom you have worked. There are several areas to be covered.

Pictures

We've all heard that a picture is worth a thousand words. In construction selling, it's true. Have with you at all times top-quality pictures that will show the prospect the caliber of work your company does. Snapshots will do, but larger pictures are more impressive.

Snapshots can be enlarged with no trouble, but there is usually a limit beyond which the pictures start to lose their clarity. The handiest size for showing detail is 8 in by 10 in, kept in an attractive folder. I suggest that for really good pictures you contract with a professional photographer. I've used professional pictures, and they are well worth the expense. The professional knows how to use light and shadow to highlight a particular detail or catch a building at the right time of day. The more impressive the pictures of your work are, the better job of selling you can do. It takes good tools to do a good job.

The secret to using pictures is not to overpower the prospect with them. After she or he has looked at four or five, it gets to be old-hat—in short, boring. Four to five should give a decent cross section of your work, which is all that is needed.

Besides their use in selling your company, pictures are great to move the job along, to get the owner thinking about the details of the new building. This use requires more than the small thin folder used for selling the company. You need to have pictures of all different types of buildings, and you should break them down into categories, such as shopping centers, auto dealerships, retail shops, office buildings, and storage warehouses.

These pictures don't have to be of your jobs. They can show anyone's jobs, although you should ask permission of the owner before you take any pictures. The same is true for the jobs you have built. I don't know of anyone who has ever objected, and it's a thoughtful courtesy.

Trade magazines are an excellent source of good pictures; don't hesitate to use them. These pictures aren't shown to the prospect unless you feel they can be helpful. For example, while looking at the company pictures, prospect X singles out one, saying that's what he or she is thinking about, but with larger plate glass and a brick front. You know you have something close in your picture file, so you get it. It doesn't have to be exactly what prospect X has in mind, only close; but it really can give the prospect a lift to see a picture of what is desired. I sold a tire store by using a metal building manufacturer's sales brochure that had a picture of just what the owner wanted. When I showed it to him, he said that was exactly what he wanted the new branch to look like. I had a signed contract one week later for a 7200 ft^2 retail tire store.

Company Brochure

I've had occasion to use an attractive, professionally done company brochure, and it can be an excellent sales tool. You're able to accomplish several things simultaneously with the company brochure. You can present the whole company image in one neat package. Pictures of past jobs along with captions giving the size and use of the buildings plus the owners' names allow you to show pictures, drop names, and provide a ready list of customer references at the same time. You have the added advantage of being able to leave the material with the prospect to study later leisurely. It also keeps your name in front of the prospect much better than a calling card would.

I've had experience with company brochures, and there's one thing not to do if the pamphlet is going to be used as a sales tool: Don't show pages of employees' pictures and lists of high-powered personnel with all their qualifications. Show buildings; that's what you're selling, not people.

I've seen a brochure full of people backfire. The prospect was scared off because he saw the company was obviously carrying a very heavy overhead load and just knew it was going to be reflected in the price he paid.

Showing Jobs

Putting the prospect in your car and showing actual buildings that your firm has constructed is another way to sell your company's ability. Be prepared for the prospect to offer some resistance to this idea. I don't know why, but it's always a big deal for prospect X to get away for a while, even to do something that seems important. Remember, if she or he does agree to be shown some jobs, it's a very good indication of interest. When you actually get X in the car, she or he becomes a captive audience; don't make the prospect sorry to be in the car with you so that he or she can hardly wait to get back to the store. Do a little talking about the buildings you are seeing, but don't overwhelm the person with construction details. The prospect couldn't care less! He or she will want to hear about how much a building costs, how so-and-so is doing since moving in, or whether that type of sidewall treatment is expensive.

Remember to do some listening; don't do all the talking. And try to show what is of interest; don't show schools when the prospect is thinking about a new warehouse. Don't keep the person away from business any longer than you said you would—be very careful about this. If the prospect would like to see something a second time or a building that might be a little out of the way, then go. And don't hurry back because you have something else to do. If I have a meeting scheduled that may go into overtime, I never commit to another appointment behind the meeting. It's a very bad selling technique to have to cut it short when she or he is interested in talking with you. If at all possible, plan your time so there will be no conflicts.

The situation with showing jobs is the same as that with pictures; jobs can be used to sell the job as well as the company. In showing a job you'll want to take the prospect on a tour inside if possible. And right here the bad jobs come back to haunt you—it would be something if prospect X wanted to see the interior of the building you did last year and the job had been unpleasant at the end! I'll never be able to say it often enough: The happy owner is the best sales agent you'll ever have.

Okay, so you have an owner who will let you come back, so you arrange a meeting to show the building. Always do this. Don't barge in; that's bad manners. Here is where the car phone can be put to good use. When you and the prospect show up, go straight to the owner and introduce the prospect. You may be surprised; chances are, the owner will take over and conduct the tour.

Since you are in the construction business, you know from firsthand experience that no job is problem-free. Even you and the owner now conducting the tour probably had a couple of problems. The funny thing is, the owner probably will not mention them. The owner wants to give the impression that she or he was smart enough not to have any problems. Besides, it's flattering to show off the building. You shouldn't let some little problem stop you from calling on an owner to show her or his building. At the same time, don't be dumb enough to ask if you know the owner will refuse. Why ask for grief?

A good salesperson lets the product speak for itself. You do the same thing by showing some of your past work.

Dropping Names

Be a name dropper. Every place has its prominent citizens. They don't necessarily have to be known to the general public, say, such as members of city council. They can be individuals who are recognized as leaders in an area familiar to the prospect. Use their names if you have done work for them. It helps put you a step ahead of the competition when the prospect knows your firm is accepted by the movers and doers in your community.

It helps as well not to hit prospect X over the head with a name. Slip it in gently, and then let the prospect pursue the subject. For example, don't say to prospect X, "I built So-and-So's new facility." That might be taken as boastful by X. Instead say, "I faced the same problem, just last year, and the situation was taken care of after we brought in the city engineer. It happened to be on the job we did for So-and-So." Now you've told the prospect that your firm has done work for a certain person in the hopes that it will be impressive; and you've done it without being too pushy. At the same time, you're telling X that you can solve problems.

Along with name dropping, be prepared to give out references. The convenient way to do this is to have reference material made up that becomes another selling tool. (See sample on page 139.) Here you list several jobs, and I would suggest no more than five or six. That's enough to get the point across without overwhelming the prospect. Along with the job names give the address, phone number, and the name of the contact person. This keeps things easy and simple for the prospect to use.

This reference material is designed to sell your company, so include your bank and a name to contact, then the insurance firm with a name, and a list of suppliers with whom you work. This reference material allows the prospect to check out your firm and determine that you are what you say you are.

There will be some work involved because everyone on this list should be contacted and approval given to use the name and company. Also, here is where you can stress a very important point that is a part of design-build construction: repeat work. Whenever possible, let your job list have repeat customers. On the sheet I'm using, all five of the projects are second and even third jobs I have done for the same business. This tells the prospect that your company is well accepted in the business community.

Reference Data

Keep the reference material to one sheet of paper with the company letterhead.

Company Letterhead

XYZ Construction Company
Address
Phone, fax

REFERENCE DATA

Note year when company started

Longevity in business looks good to the prospect. However, if it's a short time, say, 1 year or 2, I suggest this be omitted because the prospect may take note and want a firm with more experience.

Projects

List four, five, or six (no more, that's all you need) projects with phone number and party to talk with. This provides the prospect with positive evidence that your firm can do exactly what you say.

Credit

Here I suggest you list three particular areas:

1. Bank, phone number, party to talk with
2. Insurance, phone number, party to talk with
3. Suppliers and subcontractors

List five firms with names and phone numbers. The credit references let the prospect make her or his own determination that your firm is a respected member of the business community.

Let me give you an example of the reference material that I use.

BOOTH CONSTRUCTION LIMITED
Specializing in Metal Buildings

Reference Data

Company started in 1977

Recent Projects Firms that had Booth Construction back for repeat jobs:

XYZ Corporation	Phone number	Contact person
ABC Seafood Company	Phone number	Contact person
Fix It Auto Repair	Phone number	Contact person
Any Color Paint Company	Phone number	Contact person
Wood Materials Storage Company	Phone number	Contact person

Repairs Firms that needed metal building maintenance and repairs:

XYZ Packaging Company	Phone number	Contact person
ABC Truck Line	Phone number	Contact person
Building Supply Company	Phone number	Contact person

Credit

Bank	Phone number	Contact person
Insurance	Phone number	Contact person
Metal Building Supplier	Phone number	Contact person
Local Concrete Company	Phone number	Contact person
HVAC Company	Phone number	Contact person

Now this is just an example I use. I have been at it long enough that I find it very helpful to note repeat work under Recent Projects. This shows the prospect that the owners were pleased with working with me, which shows I do what I say I will do, and this is a good selling point.

If you don't have some repeat work yet. Then use the Project section to list jobs you have done. Even go so far to include a repeat job or two if you have them. In other words, do what you can to make yourself look good in this section, and apply the same method to the Repair section.

When you lay out your reference data sheet, make sure you leave some room to write in another reference company or two. And the reason for this is something I learned from experience. You are talking to prospect X about the new building he or she is interested in and how it will be used.

You know that none of the reference companies use their buildings as X will use this one. It will help you to show X that you have already worked through the problems her or his new building will bring up, so you add to the reference sheet a similar building for X to check out. This makes you look good, so if you can use this type of reference example, do so.

My last comment on your reference material is to make changes as time goes by to make your company look better.

Selling yourself and selling your company are really rolled up into one big selling process. It's very difficult to separate the two completely and concentrate on one at a time. Time is one problem. Not much time is required to sell your company; after X accepts the fact that your firm can indeed do the caliber of work needed, that's it for that subject. On the other hand, selling yourself is an ongoing process with no actual stopping point. You have to be able to jump back and forth between these two areas.

This is all fine, well, and good, you say. But how do I know when to stop this type of selling and get on with selling the job? The answer is: I don't know! Prospects are all different in some ways and alike in others. It is strictly a judgment call on the part of the construction sales representative. You might call it a feeling, a change in attitude of the prospect. The successful salesperson is constantly on the lookout for such changes, but don't expect them to hit you over the head. More than likely, it will be an innocent question. The best yardstick I can offer is interest on the prospect's part. Questions will be the indicator to watch for; and when you think the time is right, get to it!

I also recommend using the negative reference. This is not a reference who will bad-mouth you. *Negative reference* is a term I started using several years ago, the idea being a holdover from my earlier days on the road as a factory sales representative for a building materials manufacturer. The old hands told me when I started selling to always remember that a well-handled complaint will impress your dealer more than anything else.

I put this idea to work in selling construction. While I'm giving out the reference information, I can usually get a quick doubletake with, "And while you're checking, call So-and-So. We had the biggest headache on that job." The prospect will be a little surprised to have this type of reference offered. I go on to explain that my company, too, has its share of problems. The point I want to make with X is that what really counts is how the problems are handled. I make this very clear and invite X to call the negative reference and check. I've found that such an offer carries a lot of weight and truly impresses the prospect. And the negative reference has been checked by the prospect from time to time. I believe, though, the offer is proof enough to X that you intend to be straight. It sells the prospect on you and your company.

Selling the Job

The very first impression you want to give to the prospect is interest. You bear down on all the requirements and stress to X that you need all the information possible if you are going to do a good job of helping. *Help* is the key word. Don't start with, "Give me all the information you have." Instead start with, "In order to help get this project off the ground, I'll need some facts concerning first the site and then the building size." Don't throw the subject open for debate. You keep control. You ask the prospect particular questions, and you'll be pleasantly surprised how smoothly the interview will go. The prospect will be really starting to think about the new building, and the excitement will be there.

While you're working on the requirements, please remember you will not learn a thing by talking. I mention this for a special reason: As you gain more experience, you'll come to know firsthand, if you don't know already, that there are certain places during the selling process where long periods of silence develop, for example, when the prospect is thinking about whether the building should be 80-ft wide or 100 ft wide. Your questions will make X think about the building, so let him or her think. Many salespeople believe that every time a period of silence occurs, they must immediately start to chatter. Fight the urge. It's hard, I know, but it is better to learn now than the way I did—the prospect asked me to be quiet so he could think. It was a little embarrassing; however, I still sold the job.

A heavy silence will also be created when X reads your proposal. Accept it and sit there and think about the Super Bowl, or where you're going over the weekend, or anything. Think; don't talk; let the prospect read. After the reading will come the questions, and then you can talk.

It's important not to tell the prospect what she or he should have. Just record the requirements and ask questions that add to the requirements/knowledge. Keep in mind that you're trying to find out if the prospect has a good idea of just what is wanted, and at the same time work up enough data to put together a budget price. Your recommendations can come later. This method works very well with the prospect who is starting from square one.

Let's take another tack for the prospect who had plans and was stopped because of the prices that came in when the job was put out for bid. You've got a ready-made starting point—the plans. Again ask questions; display your interest and intent to help. Using the plans, you proceed to point out areas where savings can possibly be realized. Money is the attention getter, so sell savings and places where costs can be reduced. Hopefully by now you've found out what the bids were and what the prospect has established for a budget. If not, try to find out. One way is to come right out and ask. After all, the worst that can happen is that X will refuse to tell you.

You will not be refused very many times if you preface your question with the positive statement: "Prospect X, I know my firm can work up some redesign ideas to cut costs. It will be most helpful to know the budget we have to shoot for. Early on we should be able to have a good indication of how the reductions are running, and it'll save me from spinning my wheels if I know just what your budget is. I'm sure you don't want me wasting my time on a budget that will be hard to reach." You've taken a very businesslike approach, and for the most part X will respect you for it and give you the numbers you need. You may not actually need the figure to do your work, but this is a surefire method to obtain the numbers. The fact is, you'll probably really use the budget as a yardstick.

Think and Discuss Change

I've used this method many times to start a job moving along seriously. Don't be afraid to suggest a radical change to X if it will help accomplish a certain goal. An example may make what I'm saying clearer. Let's say our prospect is a car dealer, and she wants the showroom and offices in one building and her service area in another, completely separate structure. You work your budgets, and the price is too high. At this point you tell the prospect the price can be reduced if she puts everything under one roof. This is a radical change from the original requirements, but you are showing X how to reduce costs. You're doing your job, and you're helping. She may say no; and if she does, the prospect will have to revise her budget. Another change could be to relocate the building on the site or make changes in the site itself.

This tactic is used with the prospect after you've worked up a price, all the requirements have been worked out, and the price is still too high. You use the radical change right off the bat with the rejected-bid prospect; it is especially effective here. In this particular case, the preliminary work has been done for you.

When you negotiate with a prospect who has plans, always take the plans with you—as many copies as you can obtain. Sure, you're thinking, how else will you be able to work up the needed information? Think about this one, though. What you have no one else can look at! Right? So take those plans out of circulation, and keep them out as long as you can. Usually the prospect will keep a personal set, and there's nothing you can do about it. She or he probably won't let them get away, so the competition still has nothing.

You don't have to be an expert in all kinds of construction to be able to suggest radical changes. All you're trying to accomplish at the outset is to keep the project moving. Don't even try to quote prices. Explain to the prospect that you'll have to work on the details, and you'll need the plans for a short period. Later the real construction people can tell the sales agent if indeed there is a savings. However, if the sales agent does have expertise in this area, then great. It will make him or her more comfortable with this segment of the selling process, hence much more effective.

It is imperative that the metal building sales representative know something about construction in general. The very first rule for a salesperson is to know the product. Remember that the radical change is an attention getter; once in a while the prospect will buy your suggestion, but more often not. But you have now put yourself right smack in the middle of the project. You're now part of the picture and that's all you wanted. Any other pluses are a bonus.

Many times a radical-change suggestion will not be necessary. Small changes may do the job just as well, substituting one type of sidewall material for a cheaper one, for instance. Then there are the changes you recommend that will not necessarily save any money but will give the prospect a better facility for the particular use, such as a one-way mirror in the office so X can observe the sales area while working at his or her desk, yet remain private. Or maybe you could work out a better traffic flow for trucks around the warehouse. All such changes sell the prospect on your interest in the project.

There are several suggestions that I've more or less standardized and fall back on when nothing else is obvious: (1) skylights in storage areas; (2) windows in offices to allow X visual control (this is a good term to become familiar with; it's one that retail and wholesale prospects will want to talk about, and by your doing so, you show the prospect you know your business); (3) plans for future expansion; (4) the possibility that the prospect may have room for building some space to lease out (which I use to a lesser degree).

Somewhere in these four topics I can usually find a subject for discussion, and I've never had a prospect who didn't seem to appreciate the suggestion. With a little experience you'll be able to take the owner's requirements and see immediately how to come up with some good suggestions.

There is one special item that I always try to sell the prospect on considering, where I personally feel I would not be doing my job of looking out for the prospect's interest if I did not. The main theme is the use of the building as an insurance policy. I tell the prospect the advantages of designing the building to have multiple uses. For example, X wants a 60-ft by 100-ft building to operate a retail business: a 10-ft sidewall is more than high enough for this use. I then play the role of insurance agent and ask what happens if something should happen to X. At first X may seem puzzled by the question, but almost always the answer is that the building goes to X's heirs. Then I ask if the heirs will continue to run the business. Most of the time the answer is no; chances are the heirs will sell or perhaps lease the building. Then I ask the final question: Why not leave the heirs with something that's really worth something in the marketplace? Believe me, you'll get a quick response, and X will ask exactly what you're talking about.

At this point, I explain to X that a 10-ft-high building, while working very well for X's business, still limits the number of businesses that could use the structure. It would pay to consider a 14-ft-high building, which would be more versatile in use, hence more marketable. I point out that X could look at the added cost as a one-payment insurance policy. Every time I bring this up, the prospect considers the suggestion, and I would say that in many cases the owner does make the changes.

I don't like to leave the prospect thinking about not being around, so I mention it's also a good retirement plan for X. When X does decide to retire and maybe sell or lease the building, it will be more marketable and a good source of income. This statement is always received with a smile. At the same time, I increase the contract, which makes me smile.

Often the prospect will plan to build a single-use building, and the construction sales agent should stand ready to point out the pitfalls, from both a use and a financial standpoint. Bankers are a gloomy lot about lending money, and they always take the attitude that X will go under 2 weeks after moving in, leaving them stuck with the building. They frown on a single-purpose building. There'll be times when a single-purpose building is what is called for, and nothing can be done about it.

Budget Price

You've been hearing about the budget price all through this book. The budget price is just what it's called—a budget, nothing more. It's a handle for the prospect to grasp. It's necessary for the construction salesperson to understand exactly what a budget is and, even more important, to know how to use it.

There's no way to convince the prospect to move beyond a certain point until he or she gets a price. Everything X does is aimed in this direction. He or she wants to know how much the new facility is going to cost. This is even truer with a rejected-bid situation, only the main question is, How much under the low bid can you get?

The prospect gives out the requirements only to allow you to work up the price. You're not going to get down to the nitty-gritty of selling X until you present a price. Then and only then will X press on—or maybe fall over in a dead faint and forget the whole thing.

At times the budget price and the contract price are one and the same. This usually occurs with a simple uncomplicated job, where from the basic requirements you're able to quote a firm contract price. This price functions the same as a budget; it gives the prospect what she or he has to know, and in this case X is better off because of having an exact price. In truth, it doesn't eliminate the purpose of the budget; it only avoids having to work up a contract price later. As I said, this applies mainly to the simple jobs; for the more complex ones the contract price can come only after plans and specifications are prepared for pricing.

In working up the budget price, there are many ways to save time and money. Based on your own experience with past jobs and the experience of your key subcontractors, you or your estimator, if you have one, will be able to whip up a budget with no trouble.

The key to a meaningful budget is a decent preliminary drawing outlining all the requirements. A simple line sketch is all that is needed when you are talking with a prospect. As you question X and take down the requirements on your sketch, be careful to note all the details possible. This sketch need not be a superprofessional drafting job. It's easy enough for the shakiest hand to make a passable drawing when grid paper is used. The grid paper ($\frac{1}{8}$ in to the square is the best size, I think) looks good and allows you to work to scale. Often a problem will pop up when X's ideas are put down to scale, giving the salesperson a perfect opening to make a few problem-solving suggestions.

I can assure you that in making suggestions you do not have to be an architect to know what you're doing. It's really very easy; and after you've worked with several prospects, it will become even easier. With experience you should become good at it; in this area, practice makes perfect.

I do recommend sitting down with subcontractors in the plumbing, heating, air conditioning, electrical, and site work areas for a short crash course in their particular trades. It's important to know how ductwork is run to service a showroom, or how the overhead lights will have to be arranged, or the most economical way to handle three bathrooms, and when you can substitute gravel for blacktop. I've never known a subcontractor who wasn't delighted to tell someone about his or her trade. Also, when you have a special problem in a trade that takes some expertise, give it to a subcontractor. Most don't get a chance to work on something really interesting very often, and they'll jump right in.

When I was building the three skating rinks for one owner, we had a real problem with the electrical system. The owner wanted all kinds of lights over the skating surface, which could be manipulated in dozens of ways to obtain certain effects while the people were skating. In addition, everything in the entire building had to be controlled from a control center so that the manager could sit behind the console, much like a disk jockey, and have control over the rink. And the electrician had to coordinate that work with the people putting in the audio system because everything came back to the same central console.

I arranged a meeting with the electrical contractor and the owner, and the three of us sat down, told the electrician what we wanted to accomplish, and turned the project over to him. The man did a super job and really enjoyed putting the entire system together. Use your subcontractors because they can provide backup for your budget. I've already stated my feelings about shopping a subcontractor after he or she has helped you nail down a job, but it's worth repeating: Please don't do it!

After I have worked up the budget price, I add between 5 and 7 percent to the total amount to make sure I can build what the prospect wants within the price that I state. Always explain to the prospect exactly what the budget price is. It's a not-to-exceed figure to use. I always stress the point that it would be very easy to "low-ball" a figure, get X's name on the dotted line, and then start with the added costs. This statement accomplishes two things, I hope. It lets the prospect know I'm trying to do the best I can (selling myself), and it puts X on notice to watch out for this maneuver from the competition.

The way the budget price is presented is just as important as the number itself. All prospects have some figure in mind before you give them yours. You may never know what the figure is because it may be so low, as it is in most cases, that the prospect is embarrassed to mention it. The point is, X will have a figure. It will be based on what a friend built the same kind of structure for last year, or an industry figure put out by some national headquarters, or what X built the last warehouse for 10 years earlier, with X's idea of cost increase added in. Therefore, the prospect will have some idea almost every time, and chances are X will have it broken down to cost per square foot.

You've all heard the old adage "A little information can be dangerous." Well, it's never been more true than it is with the prospect with a square-foot price based on goodness knows what. The real hooker is that when X talks to other people, they won't always tell X the whole story—and this, I feel, is not done intentionally. The problem is that the owner confuses building cost and total construction cost. When asked what the building cost is, the owner will quote the building cost, but leave out the site work and all that goes along with it. The prospect then takes this number times the area, and presto, X knows what the new facility is going to cost in total.

There is another source of semifalse information for the prospect, and that is the owner who feels he or she paid too much. Rather than take a chance on looking foolish in the business community, this owner will deliberately put out a lower figure. One way the embarrassed owner can answer the question and not look dumb, he or she feels, is to give the building cost only. In any case, the prospect still ends up with a dangerous number, which the construction sales representative will have to handle.

When the salesperson presents a figure, it must be broken down into at least two segments: building cost and site cost. Never give the prospect a total turnkey price. I'm speaking from bitter experience. When I started selling construction, I worked with a turnkey price, and I was losing contracts at an alarming rate. It took me 6 months and the help of another sales agent experiencing the same problem before we were able to solve the mystery. Our competition was quoting building cost and site work separately. Many times the site cost was never priced, only the building cost. The building cost that our competition was quoting most of the time was somewhere in the ballpark the prospect was expecting, and when X compared our price to that of the competition, we were promptly forgotten.

You may ask right here, what's the big deal? Tell the prospect what your price covers, and she or he will be able to make a fair assessment. I wish it were that easy. Let me digress a moment and tell you a not-very-flattering fact about prospects and leads as well: They only hear and retain 10 percent of what you tell them!

When I follow up with someone after talking with her or him the month earlier and am told she or he has signed with a competitor, I don't mind it—that's business. But it really gets to me if the person then says with a surprised voice, "I wish I'd known you did this kind of work." The fact is, I did say so a month earlier; it just didn't register with the dummy. This to me is one of the most frustrating aspects of construction selling that sorely tests my objectivity. You have no way of knowing if you are really talking to the prospect or just speaking words that go in one ear and out the other. As you gain experience, you'll encounter this type of thing, and there's absolutely nothing you can do except to grin and bear it.

Thus, you can't explain something as important as the budget price and expect the prospect to remember how to use it. At the same time, it's confusing to the prospect to try to study the situation when he or she has to use two systems, especially when X knows nothing at all about construction. You cannot teach X to use your system when X has one she or he already understands. You have to make your information fit X's format, and do not expect the prospect to learn to use something new.

After the other sales agent and I discovered what we were doing wrong, we changed our budget price structure, and the contract-signing rate started to improve immediately. Lately I've found the building and site prices have to be separate, for with all the new restrictions imposed by city hall, it's almost impossible to come close to a site price without an approved site plan from the city officials. I imagine you have found the same thing where you operate—more and more added to site work by city hall that really drives up the cost and in some cases puts the project cost out of reach for the owner.

When you present the budget, push the building cost and stay away from the site details. Prospects have a way of thinking that site work will cost what it will cost, and there's nothing to do but pay; so capitalize on this attitude, and stress building cost and places to save. Nevertheless, chances are that the numbers you give the prospect will be greater than what X has in mind. You'll know this from a long, drawn-out whistle or in some cases from a dead silence that lasts for what must seem like a year. X is trying to absorb what you've just said, so keep quiet. The prospect will speak when his or her breath has returned.

After X has the price, you have to start getting X used to the figures. A trick I use here is to put the ball back in the prospect's court, so to speak. I ask what prices are doing in X's field. Can X buy items today for the price paid 5 or even 10 years ago? Naturally the answer will be no, and you tell X that you can't either. The two of you are in the same boat; you just sell different merchandise.

Usually at this point it's possible to obtain the prospect's budget. Simply say, "If the price seems high, tell me your budget and I'll see if it's possible to reach it and stay within your requirements. I'll have to make some changes, I'm sure, but there has to be a point we can reach to move along." Once you know what X has in mind, then you can go to work on any changes that may reduce the price.

Now X is waiting for you to come back so he or she can see if it will be possible to build. X is becoming dependent on the sales agent. The sales agent must work at increasing this dependency by handling details and problems for the prospect every possible time. I'm not going to cover the numbers that go into a budget price. We are concerned with selling, not estimating. Besides, every locale is different, and only you know what's going on where you operate. My only recommendation is to put in a little extra for contingencies.

There is one thing to guard against when you work with the budget price, and that is the prospect who wants you to break out subcontractor prices. Watch out! The prospect is shopping your prices. When this comes up, I tell the prospect I'll be glad to quote a price for the structure only, and X can contract with other trades directly. This does happen fairly frequently when you're selling metal buildings. The prospect will drop back to a structure price mostly to save money. There is one exception—when the prospect wants to handle a particular trade her- or himself. Say X's brother is a plumber, and X wants him to handle that area; then I will break out that price. That is the only time I will, though.

So beware of the prospect who wants a breakdown. Say what you will for an excuse, but don't give it. I handle this particular situation by telling the prospect that the prices are worked up on the past history of other jobs by the square-foot cost, it would take a great deal of time to work up subcontractor prices, and I would be pushed to do so at this point in the negotiations. But here you want to tell the prospect nicely that you won't do it. After all, X is a prospect, and you don't want to burn any bridges.

Let's talk about an area where the construction salesperson does not feel very comfortable: the use of drawings with the prospect. Here's how the gray areas becomes scary. The salesperson loses some control. You draw a sketch recording X's requirements, work up the price, and present it to X. At the same time it's difficult to explain exactly what the price covers without using the original sketch or maybe a more detailed one you used to take subcontractor prices. Some design-build contractors offer drawings with the proposal price. I think this is a little risky, though. In any case, the prospect will want to see what the money buys, so a drawing is always part of the budget, if for no other reason than to help explain the details. It's supposed to be a simple sales aid, nothing more.

The problem arises when the prospect asks for a copy of the drawing, for any number of legitimate reasons—to show a banker or lawyer, or maybe to talk over with a spouse. Anyway, X wants a copy of the drawing.

I know you can see the trouble that might arise if you do as asked. With the drawing X has all she or he needs to start shopping prices. The construction salesperson at this point is most vulnerable; he has done a lot of work, and the prospect can take it and go on his or her way with it.

The only defense you have is not to leave a drawing if you can get away with it; and sometimes when I feel the prospect might be a shopper, I don't even put the budget price in a typed letter. It's jotted on a scratch pad, and that's what I hand to X. If X tries to shop it, the competitor doesn't know if it's a price from a contractor or something that was made up. A price on your letterhead is proof to your competition.

Don't work under the mistaken idea that the prospect won't show your prices and drawings to a competitor. He or she will! I'm speaking from experience. I've had some of the most beautifully prepared proposals handed to me, without asking. When you're talking about large sums of money and the prospect sees an opportunity to shop a few thousand dollars off the price, X is going to do it. There's not much you can do about it except to do such a good selling job that the prospect really feels more comfortable dealing with you and will let you have the last look. For this reason the sales agent never stops making the prospect dependent on him or her.

Back to the drawing. I try to combat this situation of having to give out the drawing by using an original one while presenting my budget proposal. It doesn't have to be the one you started with; it just shouldn't be a copy. While you're explaining your price, use the drawing to point out details and always have several questions about some particular detail. If there's a change, note it on the drawing. If no change, note that also. The main purpose is to mark up the drawing.

When you've completed your meeting and are gathering your papers, take the initiative with, "X, this is the only copy of the drawing, and we've made notes all over it, so I'm going to have it redone and brought up to date." I've never had a prospect argue with that approach. But nothing is completely sure, and every now and then a prospect will ask to make a copy for his or her records.

There are still times when, having left a copy of the drawing, I walk away with a slightly uneasy feeling in my stomach. It comes with the territory, so get used to it. Do everything possible not to leave a drawing because you're helping the prospect become a little less dependent on you.

I know of design-build contractors who work up a full-blown proposal with simple plans of the proposed project and give them to the prospect as a matter of course. I don't recommend doing this sort of thing at all. It has to run into money, and the prospect is getting way too much for nothing.

Sales Agent's Exposure

This subject overlaps somewhat with the budget-price topic, so we'll jump back and forth at times. Think back to my statement that the prospect retains very little of what the salesperson says as well as X's impression of the salesperson. The only way to combat this problem is through exposure; in other words, X has to see a lot of you.

When you are working with a "normal" prospect who wants to build a normal building, say, a $300,000 contract, and there are no exceptional problems, the time span from first call to budget price should be around 2, maybe 3 weeks. I know this period will often vary, but I want to give you a feel for exposure because it's important. Why take so long to come back with the budget price, you ask? A few days should be enough. That's right; a few days are enough if you only want to drop a price on X and leave to wait for her or his phone call. But that's not what to do to sell construction successfully. The important thing here is to use the time interval to give you exposure to the prospect.

This period between the first meeting and presenting the budget price is what I call the *work-up period.* During this time you should make at least four contacts with X. Two or three should be in person and the others by phone.

It's important never to give out everything you know at the first meeting; always hold back something that you need to know. Even if you won't be able to put the price together without the data, don't be too concerned. Preparing the price should not be very time-consuming, so you'll have time to pursue exposure and put together the price.

At the same time there usually will be questions that should be put to prospect X; but if there are none, then you'd better get busy and think of some to ask. You'll have to be the judge of what to see X about in person and what can be handled over the phone. My guideline is: How important a question is it? To me a question is important if you cannot work up the price without its answer. Everything else, of course, is less important.

Please don't bother the prospect with obvious little questions that make no difference at all during the work-up period. They simply detract from the image you are trying to establish with the prospect. Very often during the work-up period you'll have to go back to the prospect for a command decision. For example, X is undecided whether to build a 50-ft-wide store or one 40 ft wide. You offer to help by stating whether there would be any savings and, if so, how much. I'm not suggesting that you price two different buildings. You provide X with just enough to make a decision, nothing more. It's really not too much of a job to determine that there would be a savings in glass. Anyway, the sales representative will have to have a meeting to present the data so the prospect can make a decision or choice. Then after the building is determined, you can go on with working up the price, at the same time making sure you give yourself as much exposure as possible.

I've found that solving small problems that really don't affect the price is an excellent way to stay in contact with X during the work-up phase. For example, is the water and sewer hookup at the back or front of the site? Or, will the prospect be able to put in a 40-ft driveway entry without special permission from the city? All these types of problems can usually be taken care of by phone, and then you have a very good reason to contact X with the information he or she is waiting for. All this activity helps sell you to the prospect.

Remember, it's very important to have all the exposure possible. You want the prospect to begin thinking of you as the contractor. During this time the sales agent can establish some rapport with X, which will go a long way when you get to the prices. It should make it a little more comfortable for all concerned.

Always have a reason to talk with the prospect. Never just stop by to pass the time of day. X doesn't have the time for this, and neither should you. Have an obviously acceptable reason for calling on X, and never say, "I was in the neighborhood and thought I'd stop by." No matter what the reason for calling on the prospect, you never want X to feel neglected in the slightest sense. You might actually be in the area, but never let X know this. You want X to think you made a special effort to talk with her or him about the project. Again you're selling yourself.

While it's most important for the construction salesperson to gain exposure, care must be taken not to overdo it. In other words, don't become a pest and start to worry the prospect. The sales agent must exercise mature judgment. I'm at a loss to pass on any meaningful guidelines. There are so many little things that make up the prospect's personality, and the sales person is the only one who knows how to handle X. By now you should have some idea of what you can and cannot do, so it's up to you to determine the best approach and to be able to keep selling yourself without pushing X's *off* button.

The sales agent must learn to be attuned to the little things that warn to back off a little. Maybe prospect X is a little short with you when you contact him, or it could be a put-off on a meeting. He may suddenly be hard to reach by phone. The sales agent has to be able to read the signs.

A salesman friend years ago told me a short story about reading signs. He said a sales agent is exactly like the Indian scout of old. Both operate mostly alone, surrounded by the unfriendlies, and the only way to survive is to be able to read the danger signs left along the trail. You know, the analogy is quite true; the successful salesperson is one who can note the signs, both good and bad, and can conduct herself or himself accordingly. There's no book you can learn this art from; only studying human nature and experience will sharpen your abilities.

Timing

Timing in any sales endeavor is critical to success. It is doubly so in construction sales. Think back to my example about the toy man who decided not to build because he was having a bad day. I'll always feel I dropped the ball with that prospect. I waited all day to take the contract to him when I should have been the first person he saw that morning after he opened his doors.

The first rule of timing is never to make a price presentation if the prospect is in a hurry. You'd better recognize the problem and tell the prospect you'll come back when he or she is not so rushed. X will only hear a portion of what you have to say under the best of conditions; if X's attention is elsewhere, he or she won't hear a thing.

Realize that when you call, the prospect may be under pressure. Pay attention when you enter the place of business. If the pressure is obvious, then get X's attention and say you'll come back or call later. Remember, you will always take second place to the business, so don't get bent out of shape when the meeting you've rearranged your schedule for is called off. Once I tried for four weeks to pick up a prospect and show him a job that he wanted to see. Every time I stopped by his place it was bedlam. I would wave and tell him I would catch him later. After a couple of times I knew the prospect was becoming embarrassed about the situation, so I went out of my way to assure him that the canceled meetings were causing me no inconvenience. He seemed to appreciate my being so understanding about his problems. And I did sell him a 4000 ft^2 building.

The second rule of timing is to try not to leave a decision to be made over the weekend. For example, when you have a price to present and you need to get X's name on the dotted line, don't do it on Friday. Invariably the prospect will take it home over the weekend, and that's when the trouble starts. X has time to talk to the brother-in-law, who always knows of another contractor who, he's sure, can better the price. I even hesitate to give the prospect a decision on Thursday; it has a way of ending up on Friday, then Monday.

A few years ago a stockbroker friend told me about some customers who played the market daily and always sold before the weekend; their theory was that the whole world can collapse on Saturday and Sunday and affect their investments. This is just a point of interest to show that construction sales agents are not the only people who recognize the negative results that the weekend can wreak on a business deal.

Momentum is as important in selling a building as it is in a football game. When the players have momentum built up, it shows in the way they march downfield, rolling over the opposition. But let them stumble one time, experience a quarterback sack or an offensive penalty, and you can almost see the momentum walk off the field. The same phenomenon applies to construction selling. There is a certain driving force that prospect X gets caught up in, which makes X want to see the new facility become a reality. This momentum will carry over after the contract is signed. Seldom does the prospect wonder whether he or she has done the right thing and try to get out of the contract. X may have passing thoughts that are kept private, but once X has signed, X is committed; and X wants to keep the ball rolling.

But let the prospect get sacked by a blitzing linebacker, and momentum will be lost. And that's what the weekend does; it allows the prospect to sit down and think about the money, the responsibilities, the problems, and, much worse, to have them pointed out by someone else. The net result is a loss of momentum. This doesn't necessarily mean the prospect will not sign the contract. Once in a while the weekend reflection period makes the conviction to do the job that much stronger, but that's the exception. Usually the salesperson will have to start all over again selling the prospect—maybe by assuring that the problems really can be solved or perhaps in the form of a change. The prospect might decide to start out with a smaller building than that originally planned. In any case, the construction sales agent should be prepared to "hold the prospect's hand" for a few days and work like the devil to get the momentum going again.

What happens when it's not possible to keep the prospect from having the proposal over the weekend, you ask? Nothing. It's a dilemma that comes with the territory. The salesperson's only defense is to time it so the prospect will get the proposal on a Monday or Tuesday. I use a tactic that will be helpful to you. I leave the final day on which I'll give the budget price to X tentative, which gives me a good reason to make that last exposure call. If you're thinking that's not much of a reason to call on X, just to say you'll see her or him on a later date, you're right. So I give the reason for the call substance: I tell the prospect one of my subcontractor prices came back too high, and I'm in the process of checking it out and will call again next Monday. That's what I call a three-in-one shot. I timed the proposal presentation for a Monday, gave myself exposure, and let X know I'm very much interested in this project by saying I'm doing my homework with the cost.

Even when you manage to present your proposal to the prospect at the beginning of the week, there's no guarantee you can have it returned to you before Friday. This is a difficult thing to control. Nevertheless, try to control it as much as possible, and then follow up to get the answer before the end of the week.

Watch your timing, and try to catch the prospect in a receptive mood. When the prospect tells you that he or she is ready to sign, drop everything and strike while the iron is hot. Don't delay any longer than is absolutely necessary.

Often you'll have to prepare a letter or even a contract for the prospect to sign. This happens when changes have been made to the budget price proposal so it can't be used; or perhaps the prospect never had a formal letter from you. In any case, you need something for X to sign. So you have it typed up at once. Pay overtime or anything else you have to, but do it. Then with papers in hand, go straight to X.

You don't want to give the prospect the impression you're falling all over yourself to get the signature. It simply doesn't look good for the professional, so you cover up the haste with looking out for X's interest. Usually negotiations are somewhat drawn out, and prices from subcontractors and suppliers can become a little shaky later on. At some time during the negotiation, you should have mentioned the importance of nailing down prices. That lays the groundwork for you to hustle and get the contract signed. You're moving fast to secure firm prices as quickly as possible, and that's why you rushed right over with the contract. The metal building gives you a perfect reason—to get the building in production schedule so that the waiting period is lessened.

Proper timing is necessary for a smooth job after the contract is signed also. When you need a decision, try to present the situation to X when the prospect is on the upswing. For instance, you would like to sell the owner on a change order that will benefit her or him but also put a little more money in your pocket—say, change the windows to insulated glass, or upgrade the heating plant to take care of future expansion. Approach the owner on the upswing, when positive events are taking place, such as the slab's being poured. My favorite time is when steel is going up and the building is taking shape. I think you could sell the owner almost anything when she or he is looking at the roofline against the sky.

Of course, you have to take into account when the work has to be accomplished and keep any changes in their proper order. But whenever possible, sell on the upswing. This also goes for bad news, which in the construction business always seems to center on time schedules.

You must move to get the prospect's name on the line, and cover your haste by saying you're saving money and/or time. Timing is extremely critical, so learn how to make it work for you.

Help with the Financing

As I said earlier, you'll be astounded at how little the prospect will know about money. Oh, sure, X knows everything there is about their particular field, which mainly boils down to buying and selling. The monetary mechanics of this process are not at all complicated when compared to financing a new facility.

The first problem facing the prospect is that bankers and mortgage people have their own jargon. So do people in every other field. People in their own field expect the outsider to understand what's being said. The problem is it's often gibberish to the prospect, who is somewhat embarrassed to admit not quite understanding all there is to know about take-out commitments, placement fees, or exactly what a construction load is and how it differs from the permanent mortgage. Or in some cases, if the prospect is going to be leasing space, there may be a rent-up platform to meet. Then you can get into interest rates, time frames, and tail-end balloon payments. And there's much more to utterly confuse the prospect, who only wants to build a new building in order to sell more paint, tires, or whatever.

Add to this all the paperwork required, and you'll think you're dealing with a government bureaucracy. Let me repeat what I said earlier: If you're not comfortable with the financing aspect of selling construction, then I suggest you take the time and effort to become knowledgeable in this area. I'm sure you know a bank or mortgage company official; ask for some help, and she or he will probably flood you with reading material. It's a very good idea to stay up to date in this field. Learn what's new and how you can use it. Keep up with interest rates and who's lending money at a given time.

Being able to talk over these matters will be invaluable in making yourself useful to the prospect, and that's the name of the game. You must be careful, though, not to talk down to X if you see an opportunity to help with the financing. You should be as comfortable with the jargon as any banker, but don't come on as one. The idea the construction sales agent wants to impart is that it's the prospect and the agent against the other "guys."

I've learned that the best way to approach the subject with the prospect is to offer to provide information. It's important to give the money lenders a professional package with all the necessary data about the project. Let's look at the *bank package.*

When she or he goes after money, the prospect is in the selling game as much as you are. X has to sell the lenders and prove at the same time that X is good investment potential. One of the most effective ways to do this is for the prospect to show the loan people a first-rate construction firm with a track record. Money lenders are cold, hard realists; the uppermost question in their minds is, How is the loan handled if something should happen to X or the business? This, of course, is only one of many questions that must be answered, but it nevertheless is a key concern. The idea is that your bank package will show the lenders exactly what their money will be used for.

When I work up a bank package, I include three parts: a cover letter explaining the requirement as given to me by X and stating the price, a description of the work to be done, and the drawings. I always point out that the price is not-to-exceed budget, but nevertheless still a budget. Then I explain that a firm contract price can't be given until architectural drawings are prepared and priced, if that is the case. In some cases the job is uncomplicated, and I can give a contract price; then I state in the letter that we are dealing with a firm contract price. In all cases, architectural drawings will have to be prepared and put on file with the bank for reference during construction.

The work description includes a specifications list detailing what material is to be used where. This list doesn't have to be a long superdetailed set of specifications. I hold mine to around two pages, sometimes three. Remember, we're dealing with lending people, not construction people, so the list should be kept simple and easy to read. They will not be interested in the size of the rebar and thickness of the concrete in the foundation, only that it will be designed by a licensed architect or engineer and constructed to the design.

For the drawings, one will be a site plan showing the building location—and I don't mean a site plan designed by an engineer that you can build from. I mean a plan showing only the size and shape, and exactly how the building is situated. The second sheet will consist of a floor plan and one or maybe two elevations. The size of the drawings is important. I know it's easy to reduce them to 8 in by 11 in for convenience, but I feel it helps put your image across when you supply a larger blueprint, and I always use blueprints, not copies from the office machine. It's possible many times to have all the information put on one page, and that is convenient. But I'm after good appearance as well with my bank package, and five or six pages plus two blueprints make for a nice hefty booklet; it looks substantial. It's important not to furnish the blueprints separately but to fold and insert them in the folder cover. This keeps the package in one part, making it handier for office use. Along with my bank package goes my offer to accompany the prospect to the bank and be on call to explain any details that may crop up.

A bank package represents time and money, and you're reluctant to do all this without a contract. Well, so am I. The only time to become involved with helping the prospect secure money is when you're protected. The way I handle this—here is where I'm getting ahead of myself—is with a contingency contract. I have a signed contract stating that if X can secure financing, I have the construction contract. I'm willing to gamble a little with the prospect, knowing that by doing all I can to help obtain the money I'm helping myself nail down a contract. I'll cover this again in Chap. 10 and go over another method to use for protection.

I do admit I don't go this far if the prospect is the least bit shaky. But by the time you get to this point in the negotiations, you should have a good read on just what type of prospect you're dealing with.

Another problem to be aware of is that it takes time to obtain a loan. Money people never seem to be in a hurry, and you have to wait for them to act. This time lag causes a loss of momentum. It's to the construction sales agent's benefit to have one or two reasons to see the prospect during this period; it will help keep interest and momentum from lagging too badly. Finally, be prepared to lead the prospect step by step through the financial woods.

Build/Lease Options

From time to time you'll come across a prospect who has all the qualifications for a profitable job except that she or he doesn't want to be tied up with mortgages and the spending of capital, which go along with financing a project. The new facility will mean more inventory, and more inventory means the prospect needs more money.

Usually you'll find the prospect in a dilemma as to how to go about accomplishing this goal. The problem is that X doesn't just want a larger building; X wants one in which to work from efficiently, and in a location that's convenient to the trading area. There will probably be plenty of places to rent, but most of the time X will have to take something less than desirable and lock in for years on the lease. The prospect can't see her- or himself working and buying the building for the landlord.

The prospect ends up with a special set of requirements: having a building constructed for the prospect's particular needs on the site desired and paying rent for a few years with the opportunity to buy the building. Who can arrange all this, plus work out all the details that go with a commercial project? Our prospect has a hard order to fill. X knows general contractors will build, architects will draw plans, bankers will finance, and commercial real estate people will work out the lease details; but X is in the wholesale food business and wouldn't have the time to coordinate all these people to make the project fly even if X knew how.

Then in walks the construction sales representative, who explains to X that she'll take care of everything; all X has to do is sell food products. What do you think X's reaction to the saleswoman will be? You're right; X will welcome her with open arms.

If you're thinking this sounds too good to be true, you're dead right! The time involved is a killer. The salesperson is responsible for everything, and there won't be enough hours in the day, you'll think. The rewards will be there, though, believe me; I speak from experience: two 18,000 ft^2 skating rinks, one 30,000 ft^2 food warehouse, and one 20,000 ft^2 moving and storage warehouse. In all these cases, the prospect wanted to keep his or her working capital intact and get into a new facility, and the lease design-build plan made it possible.

This type of prospect is located and sold exactly as any other prospect with one important exception—there will be a third party involved. What used to be a two-way, one-on-one situation now becomes a building triangle of tenant, owner, and general contractor.

As I said, you follow all the guidelines with the prospect. You still need the requirements, a budget price, and as many details worked out as possible, because lease design-build will necessitate the construction sales agent's selling the job twice—first to the prospect and then to the future owner. It's impossible to do so without doing your homework with the prospect. This provides the information needed to show the future owner how the numbers work out.

Along with the construction data, you have to arrange for the tenant and the landlord to meet and negotiate the lease terms. Don't assume you'll be present for this particular meeting; you may not be welcome. I always offer to attend the meeting, and several times I've been present; also I have been politely told my presence was not needed.

There is one very important point that the construction sales agent must always remember: When it comes to the bottom line, you are working for the person who signs the check. This is sometimes difficult to do because you are in essence serving two masters. This arrangement at some time will lead to a disagreement; it may be just an insignificant trifle or a knock-down, drag-out fight. When you find yourself in this predicament, remember whom you're working for. I'm bringing this up because I've learned that you seem to be on the prospect or tenant's side. I don't know why, except that you have built up a rapport with X, and maybe there is some empathy on your part.

After you sell the job or idea, you have to locate an owner. There are two main possibilities: First, you and/or your company can become the landlord; second, you can get one or more outsiders to be the owner.

Let's look at the first suggestion. If you and your company are in the market for some excellent real estate investments, then this type of enterprise can be very attractive. The property should appreciate with time, and hopefully there will be some cash flow as well as tax incentives. There is one place to be very careful—ask for and expect complete up-to-date financial information from the prospect or tenant. The owner who commits to the tenant had better be sure the tenant can meet the monthly payments.

The second suggestion takes more of the construction sales agent's time, so you'd better be prepared. At this point, "digging up" an owner is like digging up a lead. You have to know where to go and what to do, with one large exception, which is that you'll probably be dealing with some "serious players" in the investment circles in your business community. Locating these people so you may call on them will require a combination of many methods. Of course, the personal contact is one of the best ways. The method is up to you more or less, but one thing is a must—you have to go out and ask; they're not going to knock your door down.

A good starting point in looking for an owner is to contact people in the real estate business who handle real estate developments such as shopping centers, office buildings, and warehouses. If you bring them a tenant, they may be able to supply an owner in return. One word of caution—security; be sure you don't put your tenant on the street to be scooped up by someone else. Until you know exactly what's going on, play your cards close to the chest.

When you're dealing with a third-party owner, your situation is one-on-one, and you have a certain amount of control. It's also much easier to communicate and know where you stand. On the other hand, when you are dealing with a syndicate, it's no longer one-on-one, but one against a bunch. There will usually be one who brings up the question of construction cost and asks how they can know the construction costs are competitive. I know what you're thinking: You take the deal to them, explain the numbers that have a profit for them, and someone asks about construction costs. You hand them a deal on a platter, and someone is worried that you might be making some money out of it! Face it; there will be people like that. I've often wondered how such a person would react to my sending him or her a specifications sheet for a physical exam or detailed legal work and asking for a quote. I'm mentioning this so you can cover yourself at the beginning.

As soon as I meet with the group, I tell them, in a nice way of course, that I'm there talking to them for the money I expect to make. I'm not putting in all the time and effort that a project of this caliber requires in order not to end up with the construction contract. Then I go on and explain the obvious: Their main interest should be in their return. If this project clicks, and we all come to terms, I want it clearly understood that I build the building! This will take care of the price-conscious people, for they are listening to you for the same reason you are there—money. I firmly believe the investors understand, appreciate, and respect this approach. The trick is not to handle yourself in a manner that pushes their off buttons. Remember, you're selling! And the sad fact may be that you need them more than they need you. With the money they have, offers are coming to them all the time.

At this point the construction sales agent enters a gray area where he or she is exposed and vulnerable. So far everything with the investors has been verbal, and to provide them with the data necessary to form an opinion, you have to disclose all the information. You're hanging on a limb, and there's little you can do about it except push as hard as possible to put the project together. One advantage of working with a real estate broker who understands business ventures of this sort is that the broker should qualify the group before you have the meeting, to lay everything out for them. Knowing they are interested prior to your meeting will increase your chance of a sale. At least you won't be shouting from the rooftops for an owner.

There is another very good source of lease design-build ownership that should be checked out: firms that have money to invest and are looking for the chance to become the owner in a building triangle. I've personally worked with this type of firm, and their sole responsibility is to locate investment ventures and become the owner. It's a pleasure to work with them because they are knowledgeable about the field and understand the contractor's position, especially the fact that he or she intends to be the one to build. If you are not familiar with this kind of investing company, then check with people you know in the financial field—mortgage officials, bankers, and don't forget the stockbrokers. They should be able to help.

Let's assume the group of investors buys the package. You are now going to work with a group. They are busy in their own fields, and I've found that they don't want to be involved in the daily details. Sometimes one member will be the spokesperson. More often the real estate agent will act as the contact person for the contractor and the new tenant. I've found this to be a very good arrangement, mainly because you can usually get in touch with the real estate agent with very little hassle.

It's important to show the outside owners that you're just as much a professional as they are. Take the same pains to put together your presentation as you would a bank package. The bank package can be a good outline to follow, with the investment numbers added; always have a copy for each party who will be present.

Your biggest problem will be time. You and the prospect are ready to get the show on the road, and the group isn't able to meet for 2 weeks. With one owner it's not too much of a problem, but with a syndicate it seems that one is always planning to be out of town when you want to hold the meeting. It will be a trying time, and all I can say is to keep swinging. If time does start to drag out, push a little to get it moving; for the longer it takes, the more apt the prospect or tenant is to start having second thoughts, and loss of momentum can set in.

The build-lease package offers a very good source of profitable business for the construction sales representative who makes the effort to learn about all the parts that constitute the building triangle. All it takes is some hard work and time. Ironically, selling the prospect is the easiest part of the triangle; locating and selling the owner or owners is the hardest part—and under no conditions should you ever forget that this owner is the party signing the checks.

Selling Parts of the Project

Whatever the reason, the construction salesperson sometimes will be faced with a price that is too high. The prospect cannot or will not spend the amount required to give what X wants. Operating under the principle that something is better than nothing, the construction sales agent has to be able to show X how to save money—and one of the best ways is to take the project apart, so to speak.

The first tactic is to suggest that the *prospect* contract out the site work. In your budget price this is separate, but the sales agent is planning to do the work. Explain to prospect X that with a site plan X can contract directly with the site contractor, and you'll be glad to supply some names. Often this offer is enough to eliminate the overrun. But then there will be the prospect who needs to have the price drastically reduced; when faced with this situation, I start selling the structure only. The metal building is a very simple matter to quote; and if you get the job, it's a fast off-and-on project. When you offer these options to the prospect, you must stay in control; the best way to do so is with the plans. You explain to X that along with your portion of the work, you will furnish all the necessary plans for X to contract directly with all the various subcontractors. In addition, you'll be glad to recommend certain dependable firms to contact.

My experience has been that the prospect will use the people the sales representative recommends. This allows the salesperson to do a good turn for the subcontractor, which will pay dividends down the road. Of course, if the subcontractors you recommend are the ones you were planning to use, then so much the better. They get the work, and you have a rapport with the subcontractors that makes for better job relations during construction.

Let's return to the budget price for the moment. We discussed the changes in price that you had to supply for the prospect and why those changes are part of the picture. I'm bringing up this subject again to stress the point that many times the changes that necessitate new prices can and will be initiated by the sales agent. When you start dismantling a budget price or selling up, you're giving the prospect some new things to think about, and cost is very much a part of this package. So be prepared to do your homework and make the price changes; also offer new ideas along with the cost.

Overcoming the Desire to Bid the Project

Early on when you start selling metal buildings, you'll be faced with the prospect who is not sold on negotiating the job. X wants to put it out for bid to be sure of getting the low price. This objection is overcome by complete salesmanship.

The way I handle this problem is to attack rather than wait and end up on the defense. I tell the prospect at the outset all the advantages of negotiated design-build over the bid route. The key point is design-build. I push the fact that prospect X is going to get exactly what X asks for designed for X's particular needs. This approach appeals to the ego to a certain extent and lets X feel that she or he is really making things happen.

The first advantage is money. Some will say it is the only advantage, and maybe they are correct. I tell prospect X that if X hires an architect and bids the plans, X will supposedly know the bottom price. The fallacy here is that X only knows the bottom line for that set of plans, and those plans may call for materials that are expensive and not necessary for a functional building.

I keep stressing that with the design-build concept the prospect will be able to know what the planned project will cost. If X goes out to bid, X will have to spend a large sum of money to have the plans prepared; then there is no assurance the price will be in the range desired. At the same time, the prospect has very little control over what materials are used, and this has a direct bearing on the job cost. I ask such questions as; what do you do if the bid comes in too high and you feel it's impossible to build?

Then I go into my sales pitch. I stress control at all steps of the project. I'll supply the prospect with necessary information as to budget prices. Then the architect who is brought in will be working for the contractor, not the prospect. This working arrangement allows a give-and-take atmosphere between contractor and architect that will enable the prospect to receive the best job possible for the money spent. The contractor and architect work out a set of guidelines. The architect then has a set of guidelines to follow when drawing up the project. The architect is paid to provide a technical expertise, but the design work has already been laid out by X. Both you and the architect are working hand-in-glove to please the prospect. The trick is to give X what she or he wants, and you as the general contractor can see that this happens if the architect works with you.

The second advantage I can offer is help. Numerous problems will arise, and I will be in a position to take the load off X. I talk about helping with the finances, supplying a bank package, and being present at loan meetings. I push past performance on similar jobs (that is, if possible) and how X can profit from this experience.

The third advantage is turning to my benefit the widely held belief that one can never get anything done when dealing with contractors. I hit hard that there is one overall person in charge and everything goes through him. I call this *single-source responsibility,* and I keep driving home how it helps the prospect.

The last advantage I push is time savings. This should be important to prospects because they don't move until forced to, and then they are most likely operating with a deadline. I explain how time-consuming it is to have plans prepared and then put on the bid market. The prospect is always surprised to learn that an architect generally cannot turn around plans in a matter of days or even a couple of weeks. Then when I tell X it will be 4 to 6 or maybe 8 weeks before the bid prices will be back in X's hands under ideal conditions, X becomes concerned. The basic problem is that the prospect has no yardstick to measure performance. X knows how it's done in X's business and sees no reason why other people can't get results as X can. Bear in mind how retailers, wholesalers, and to a certain extent industrial businesspeople operate. In a very broad sense, for these people to accomplish something, they simply take an item off a shelf and send it to the customer. For many, the time it takes to crank up a construction project is incredible.

On top of this time lag within the construction industry, there's the frustrating tangle of red tape associated with having site plans and drawings approved in order to secure the building permits. I can speak only for my working area, but every year another regulation is added that makes getting the building permit more time-consuming. I'm sure you are finding the same situation where you work.

By now the prospect is concerned about ending up in a time crunch. At this point I put all these negatives to work for me. I tell the prospect that the architect with whom I'm associated turns out plans in a fairly short time, usually within 2 weeks. There are two reasons for this, which I explain to the prospect. First, the details are all worked out. The architect takes the requirements and turns the information into plans. Second, I'm a steady repeat client for the architect and thus get preferred service. It just won't do to tell X you're faster. Talk is cheap, so explain how you're faster. After all, prospect X is no dummy; don't treat X like one. If X ever thinks you are doing that, you can forget about the contract.

Next I stress how I handle the time problem at city hall. Armed with the prospect's requirements and site information, I visit the various city departments concerned and determine exactly what they expect in the plans. This really is a time saver, since revising plans after they've been turned down seems to take twice as long as it should. I especially do my homework on the site plan, for two reasons: time and the fact that the site cost is nothing but a grab in the air until you have some concrete facts to go on, and the city can provide them.

I'm selling myself to prospect X and at the same time showing X how I can get the job moving and keep it moving as well as supplying X with needed information. When I go to city hall, it's because I need help in some form. I simply ask for the help of the person with whom I'm meeting. Once in a while it's a sticky problem of interpreting the codes in the gray areas. I've never had a building official fail to try to be as helpful as possible when I've asked for help with information in solving a problem.

Back to our subject of selling the prospect on negotiating instead of bidding. After explaining how I can save time at city hall, I start to push saving actual construction time, mainly during the start-up period. When the job is bid, each part follows the one before it in a step-by-step process. Very little is done simultaneously; thus, a great way of saving time is not utilized.

The example I use is the buying of the metal building. There is a start-up time period; so, I explain to X, while we are waiting for the permits to clear, I put the building on order. This allows the order to be in the production schedule, yet it can be withdrawn before a certain time if a problem arises and delays the project. In the case of a metal building, the building can be ordered before all the interior details are worked out, thus saving time. Impress on prospect X that you can save time; and if X is under the gun, that will be more important to X.

Finally, I give the prospect my most important reason. I can save X money! Then I back up the statement. By knowing what construction materials cost almost on a daily basis, I'm in a position to recommend to my architect more economical ones. Also, my past experience on similar jobs will allow me to know what materials will give the best service for the money spent.

I know prospect X has some idea of what X wants to pay, so I push the point that working to X's budget and saving time are more important than saving a little through bidding. Overcoming the prospect's desire to bid requires the sum total of many suggestions, with the emphasis on saving money and time. Quite often you will be successful; then again, if you're not, you can always bid the job and add a redesign proposal that the prospect may find interesting.

Combating Competition

So far you have been in the front-line trenches of construction selling. But how do you handle yourself when the enemy overruns your position and you end up in hand-to-hand combat?

You'll always have competition. Without competition we would all grow lazy. Competition ensures that this doesn't happen; it keeps us sharp. So instead of worrying about competition, welcome it—it's the force that makes the wheels turn faster, more smoothly, and for less.

When you are selling prospect X on not bidding the job, the situation in which X is taking proposals from more than one design-build contractor can help to satisfy X's intention to check on the price and get a handle for the project from the marketplace. In short, X is sort of bidding the job. This situation goes a long way in convincing the prospect not to go out for a formal bid. It is the only case I know of where competition makes your job easier.

Even though other proposals give X a good idea of what the job should cost, X is still talking to your competitors. This is your next hurdle to overcome, and there's only one way: Outsell your competition! You do this by outworking them, taking on and solving more problems, and outthinking them.

The first thing you don't do is bad-mouth your competition. A professional sales agent never does this. When you do, you are in effect saying the prospect is dumb and stupid for dealing with such a no-good so-and-so. Then guess what happens? Bang! goes X's off button, and you might as well pack up and leave.

My standard reply when a competitor's name comes up is that it is a good firm and a good competitor. That's all. I'm not going to knock them, but I'm certainly not going to blow their horn. The very first point I try to make with prospect X concerning the competition is to get X thinking about who's calling on X. For example, let's say I'm from a small to medium-sized firm, and my closest competition is a large outfit. Then I stress to X the good points about a small firm. I have less overhead; therefore, I am more competitive on price, with more personal supervision and interest in the project. I'm trying to have the prospect think how these positive qualities look when applied to the larger company. I want the prospect to start making comparisons between me and the others. Hopefully in this particular case X will realize the competition is indeed a large outfit, with all that overhead X will be helping to finance, and maybe X will be concerned about not getting the individual attention X feels he or she should get.

This technique also can be used in reverse. If you're a large company, stress your good points, such as experienced workers, volume buying of materials, and better organization for faster construction. You, as the construction sales agent, will have to pitch your sales approach to bring out all the positive points. Then leave it up to X to draw her or his own conclusions. In my opinion, the size of your prospect's business has nothing to do with stressing small or large construction firms. The small businessperson may want a small contractor; then again, she or he may like the idea of the largest outfit in town doing the work. There's really no way to tell.

Hard work will overcome the competition because they may not be willing to put as much into the selling as you do. If I had to pick an area where construction salespeople drop the ball, it would be follow-up. I have ended up with a number of jobs not because of anything great I did; I just kept going back and trying to make the prospect as dependent on me as possible. You might say that I simply outlasted the competition.

So many construction companies make one or maybe two passes at prospect X then leave X to call if X wants to do business. Think back to your being on the buying end. When everything is equal—money, service, and so on—how do you choose? Most likely it's the same way I do; you choose the person who shows she or he really wants the order. So do unto your prospects as you want done unto you. It's a sure way to beat the competition.

I've also found from experience that after the prospect receives the first prices and starts working with me, there'll be some changes that will usually add to the price, and these changes are seldom questioned. So I have a chance to improve my profit picture by a small margin.

Now let's look at what you do when you get into selling the prospect late. Your competition has already been working with X. The very first thing you set out to do, after establishing the facts pertinent to the project, is to muddy the water. Your competition has a headstart on selling, and you need time to catch up, so you slow everything down by giving the prospect something else to think about. Let's say X plans to build a 100-ft by 100-ft warehouse; so you work up the price and get back to X in a few days. During this time you use the regular ways I've gone over to gain exposure while the prospect is waiting for your price. You have probably stopped the job or at least slowed it down. Then when you do present your price to X, you throw in a hooker along with your presentation. In this case you say you'd like for X to give some thought to maybe changing the shape of the building from 100 ft by 100 ft to 50 ft by 200 ft. You explain it will save some money, and the building will be easy to divide if X ever decides to lease part or all of it.

Chances are, the prospect will at least think about it. X may even call in the competition to make changes. And even if X decides against your suggestion, when the job gets back on track, you're in the running. In a case where you're coming in late, pay attention when you're working on the requirements, with the idea of picking out some point that you can use to muddy the water.

There's no guarantee this tactic will work all the time. The prospect may dismiss your suggestion with a shake of the head and push on. The prospect at this point should know who you are. This at least gives you a fighting chance. Your best defense is to take the offense. Attack your competition. Do your best to impress the prospect with your special knowledge of construction. Show X a better way to get the job done. Don't be afraid to make suggestions. This is the second area where construction sales agents fall down on the job. They take the requirements as given and work from there, without a single thought to improving what X is starting with. Nothing else impresses the prospect as much as showing X a definite improvement, especially if it eliminates a future problem.

What you're doing here is outthinking your competition. A food distributor I built for is a perfect example. Owner X wanted to build a new warehouse, and one of X's requirements was to be able to back 10 small delivery trucks into the building and leave them overnight. Early in the morning a crew would come in to load, or sometimes it was done at night. In any case, the trucks had to spend the night in the building. This also made the trucks secure from vandalism and thievery.

When X talked with me and my competition, X specified 10 overhead doors on the back of the building and explained how X intended to use them. Right away I could see that 10 overhead doors were going to be 10 headaches, and expensive ones at that. With that many doors, some were going to get banged up and cause inconvenience all the time. Two weeks later I made my presentation and received no indication at all of what X was thinking. I suspected X had heard just about the same from all involved. X could have picked any one of us out of a hat; all X had to do was make up her or his mind. I didn't like the odds, so I spent one full day trying to come up with something different that would improve the job in some way and make me stand out. I kept going back to all those overhead doors. That was a potential problem area, but how could I capitalize on it?

That night at home it hit me: Increase the building length by 15 ft and put one overhead door in each corner at the back of the building, but located in the sidewall instead of the end wall as X had requested. This would allow the trucks to be driven into the building and backed up to their loading areas. So now we had 2 overhead doors in place of the original 10, but at the same time we had a bigger building. Cost now became the factor to consider. The next day I went right to work, and to my pleasant surprise and relief I found that the price to X would not change. It was close to a washout because those 10 doors cost a lot of money.

I called X and asked for a meeting the following day, explaining that I wanted to offer a suggestion that would solve a future problem. I then spent the rest of that day working on a drawing to illustrate my idea. I even made several scale templates to represent delivery trucks. I must admit that I puzzled X when I called that afternoon, asking for the overall size of one of the trucks. X didn't know, but found out and called me back. Needless to say, I had X's interest by now.

Right on time I walked into X's office, unrolled my drawing, and described how my ideas would help X without a cost increase. To prove that the new layout could work, I used the templates to show there would be ample space to jockey the trucks about. After hearing me out, X led me out back, and we marked off the approximate distances of my proposed driving area. Then X had a driver move a truck around the marked-off area. There was plenty of room, and I let out a silent sigh of relief. I had the job signed up before lunch.

There's no way I can give you a list of guidelines to follow that will enable you to outthink the competition. You just have to do it yourself; but if you make a point of trying, I'm sure you'll be pleased with the results.

I've often heard it said that timing is everything. I agree, it is, and very much so in construction selling. We touched on timing a couple of pages back. And I would like to bring up a special use of timing here. The case in question is the timing involved in trying to slow down a job so you can catch up with your competition. After you meet with the prospect, don't rush back the next day. Drag things out if you can possibly do so, but take care not to overdo it. The reasoning here is that a construction salesperson builds up momentum along with the prospect, and you want to insert a little slowdown into this momentum. The prospect who is serious will never really lose momentum, but it's amazing to me how people selling construction grow cold on a prospect when there's a hitch. I think this goes back to my number one peeve, no follow-up. Timing is more important when you are muddying the water, but it must be handled with finesse!

A good way to use timing to combat competition is to be the last sales agent to make a presentation. If X is taking proposals from three design-build companies and you are first or second, there is no way you can expect to get a decision from the prospect until the third proposal is in. In a small way the third sales representative controls the time when the decision will be made. The first two sales agents are being forced to conform to the third person's timetable. The contacts you maintain during the work-up period can be most helpful with timing your proposal. It's easy during a conversation to drop some hints and find out when the competition is turning over its proposal. Ferret out such important information and use it to make sure you are last.

Keep in mind what we talked about earlier about presenting your proposal. Try to do that near the beginning of the week. So now you have two timing guidelines to follow: Be last, and do it early in the week. When you're last, X can't discuss your price with the competition. It will be impossible to handle this situation as you would prefer every time. You can only try, but with a little experience, it does go more smoothly.

My last comment about the competition is that just about everybody is a nice person like you, trying to make a living. As in most business areas, everyone knows everybody else and has a fair idea of what's going on; thus, it makes sense to be on pleasant terms with your competition. This can be very useful to you when selling construction. The one kind of prospect that the construction sales agent has no defense against is the joker who is using you for a source of information and drawings if he or she can get them. This is the gray area where your back is hanging in a sling, and most of the time you'll never know about the con artist until it's too late. I know it comes with the territory. But there's nothing that says you can't use all your available information to protect yourself. And one of the best methods of self-protection is to be on friendly terms with your competition and to compare notes.

I'm not advocating telling your competition everything you're doing. I'm talking about the case where it's common knowledge that you and your competition are attempting to sell a prospect, and then you start to become slightly uneasy with the way things are going. You begin to smell a rat; it may be little things that are difficult to put your finger on, or it may be something like the prospect showing you a copy of another salesperson's proposal and sketch. If X does it with your competitor's proposal, then X will do the same with yours. When such suspicious events take place, you may decide to call your competition and compare notes. A friendly phone call may save both of you some grief.

Don't spend your time worrying about the competition; spend it thinking about how you're going to sell the job. When you talk to a construction sales agent who does nothing but gripe about the competition and how they always have a lower price, then I'm willing to bet he or she is not very successful.

Price

I waited until now to discuss price. It's too easy to become dependent on price. When you do, you stop selling, and you're back right where you started—on the bid market.

I'm the first to say the price is extremely important; and no matter how great a salesperson you are, you still won't be able to sell any price you want to put on the job. If you are mainly interested in being low, then you're not going to be much of a construction sales agent. An order taker, yes, but a sales representative, no!

Out of all the jobs I've sold, I would say that 75 percent have not been at the lowest price. Nothing else means as much to me in selling construction as hearing from the prospect, "You were a little bit higher than the other person, but I know I'm going to get a better job and that's important to me." It's happened a number of times. When an owner tells me that, I make sure to give service that he or she never thought possible.

When the selling process deteriorates into a price war, I try to stress the important things that come with my higher price. I offer to make my price more competitive, but I tell prospect X in plain language X is not going to have this or that service. X is going to get exactly what X pays for, nothing else. Since X is a businessperson, this statement will make X think. There will be times when price is the only consideration; and if the job is worth it in your estimation, then go after it. Don't get angry when the prospect insists on making a decision on price alone—it's her or his money.

It's important to keep in mind that the prospective projects are all different in some ways. Each one will have to be handled on an individual basis, and the construction sales agent will have to take a flexible approach. The best guideline is to do what has to be done to get the job, if you want it. So don't discount price altogether, but don't make price the only item on which you build your presentation. The truly good salesperson talks about price last, after covering all the things that price includes.

I like to bring into my proposal presentation a thought-provoking idea, and I bring it up before we get to the bottom line: *Cheap* and *economical* do not mean the same thing. What I'm trying to do is have the prospect think about getting more for his or her money. I impress on X that I'm thinking about *economy,* and not the lowest price. I want that prospect to understand that this was the guideline I used when I put together the proposal. After explaining this, I then say that if X is really thinking about a cheap building, then I'll be glad to redo the proposal with that fact in mind. Nine times out of ten, the prospect wants an economical job, not a stripped-down one.

I've found this to be a very good tactic; use it and I'm sure you'll see what I mean.

I think the best way to sum up this section on selling is to say that service is what you're offering the prospect. You set out at the very beginning to impress on X that you want the opportunity to work for X. Service will count because you should be offering not the low price but a fair price and service. When all else is equal, chances are, the prospect will decide who does the job by the service X can expect from the contractor. When you get right down to it, service is all you really have to sell—so stress it, push it, and try your best to make the prospect understand exactly what it is you can do for her or him.

The next point I want to emphasize is follow-up. Contractors have a bad reputation when it comes to getting things done. We are always late with the work and hard to talk to when X wants to know why. Nothing ever seems to be done when the contractor says it will be. Many people do think this way about contractors, so when you follow up and do what you said you were going to do, the prospect has to be pleasantly surprised. Following up will earn you more points with the prospect than anything else you can do.

I'm talking from personal experience. I can show four buildings that I sold where the owners mentioned to me that I was the only contractor who followed up. The others didn't bother to follow up after their initial interviews or didn't even return the owner's phone call. It's easy to see how the public forms its opinion about contractors. Granted, such contractors are probably too busy to get involved. They should tell the prospect so, not ignore her or him. Follow up—it sells you and the job.

When the construction salesperson enters the selling phase, it will not be a step-by-step process as I have detailed it in this chapter. Please note that you can't sell construction by the numbers. It's necessary to be familiar with all the parts that make up selling. Then you use what will obtain results at the time you feel it is right.

The sales agent has to be flexible. When you are talking to a prospect, financing may come up before you have a chance to talk about your company references. The prospect may want a ballpark figure immediately. Be prepared for anything, and let nothing upset you. The only way that prospect will know you're a pro is if you act as one.

NOTES

NOTES

10 Contractual Procedures

After working as hard as possible, you hit pay dirt when the prospect agrees to contract with you. The question you ask yourself at this point is: Just how do I go about the necessary paperwork? Verbal agreements are fine in some cases, but not in construction sales; it's good business sense to have everything on paper and signed by all parties.

Letter of Intent

The signing-up paperwork is somewhat different from, and much more involved than, the low-bid contract. The reason is that for much of the time the sales agent is working on the front end. You have a certain amount of work to do before you can determine the actual contract price. At the same time, you don't want to be out on a limb and spend money and time without having something in writing. The answer for this interim period is a contract, in other words, a letter of intent. (See sample on page 176.)

The letter of intent is exactly what the name implies: The prospect intends to sign a construction contract after certain conditions are met by the construction company. It gives you written protection so that time and money can be spent. You will truly begin to spend money. Your architect will have to be called in as well as surveyors and site engineers. Plus the metal building is involved and time is a factor, so the building may be placed on order. Thus, it's absolutely necessary that you be protected by something in writing.

The sample letter of intent is the type I prefer to use. Note the letter is written to me by X, it is on X's letterhead, and it outlines X's prerogatives as the owner. Each party knows where she or he stands and what is expected.

In my opinion, the letter carries more weight when it is on X's company letterhead. There can be no doubt X means what X is saying when X

LETTER OF INTENT

XYZ CAR COMPANY

Booth Construction Ltd.

This letter of intent is to give Booth Construction Ltd. the go-ahead to work up a firm contract price to build our new 6000 ft^2 facility. The cost will not exceed the budget estimate of $275,000.00 unless changes are made.

When the contract price is established, Booth Construction Ltd. and XYZ Car Company will enter into a construction contract.

Note also this contract hinges on our property's being rezoned to allow the building to be constructed. If rezoning cannot be done, then this project will not be signed into a construction contract, and XYZ Car Company will pay the cost of the required drawings.

XYZ Car Company

takes the trouble to write a letter instead of signing the contractor's form. I know this is a small point, but it may help; and I know it will never hurt your case if a hassle develops and a third party has to be called in. Also, fax material only when you have to, not because of convenience. Again, personal contact keeps you up front.

Every time I've used a letter of intent, the prospect has asked that I compose the contents. After X checks it over and finds it suitable, the letter is typed and signed by X. I always take the original for my file. The prospect is aware of the letter of intent before we get to that stage with the project because during the selling process I've explained how we handle the paperwork.

The construction salesperson will encounter all kinds of conditions that will have to be put into the letter of intent. Do anything reasonable to sign up the prospect. You have to nail down the job and get it out of the marketplace as quickly as possible. I've seen very few reasonable conditions that couldn't be worked out in some kind of acceptable manner, so put all the conditions you want to in the letter of intent. Notice I said *reasonable.* As you gain experience, you'll come to realize what is reasonable and what is downright foolish. Obtaining financing is a reasonable condition. Obtaining financing at 8 percent interest is not.

This is a sample of the letter of intent I use. It's the starting point when you need more time and effort in pricing the job. This gets the project off the street so you don't mind putting some time into it.

Please note, there can be many different reasons put in the letter that will stop it from going to completion, but that's because you are at the very front end of this job, and many of the problems have not been addressed yet. So accept what is necessary to get the job moving.

Plus, if you have to have some plans drawn, make sure you note if the job does not go through, you will be reimbursed for providing the plans.

I still prefer to use the private letter for the letter of intent even though I have seen paperwork forms put together as a letter of intent. I am talking about forms where blanks are filled in with the necessary information. My reason is the private letter keeps the one-on-one business relationship up front, and the future owner likes this.

Even if there is more than one condition, use the letter, and don't hesitate to put all the conditions in it as long as you consider them reasonable. The prime thing is to get prospect X to make up X's mind and commit to you. Once this is done, the prospect becomes the owner, and the pure selling phase is over. You are now the owner's contractor, and your main job is to get busy with clearing up or helping the owner to clear up the conditions.

The most frequently used condition is financing, followed by rezoning and permits. Of course, the one condition that will always be in the letter of intent is to bring the contract in close to the budget price. Make sure you understand exactly what this condition entails. Know that the prospect is seeking financing at the going rate, not some ridiculous figure that will be almost impossible to find. If rezoning is a condition, then do your homework and find out if you're going through an exercise in futility. Don't agree to a condition you know will likely not be met. Every once in a while you'll work with a prospect who thinks he or she has nothing to lose by letting you try to put the project together, although X knows there is an insurmountable problem. Watch out for such ringers because they can really cost you. Chances are, if you do a good job qualifying the prospect, you'll know about the problems before you come to the sign-up phase. Nothing is guaranteed, though, so keep your wits about you.

The letter of intent is not worth the paper it's written on, I've had people tell me. You may be wondering the same thing about now. In my opinion, the letter is worth a great deal; it is a business document signed by prospect X stating what X intends to do if certain conditions are fulfilled. What the opponents object to is that the prospect has an out. If X gets cold feet, X just tells the sales agent that he cannot obtain financing, and the deal is off. They're right about this.

But then what do you do if, after you sign a contract, it becomes known that the financing has fallen through and the owner can't pay you? What good is your contract at that point? My attorney once told me contracts are only as good as the parties involved. If the prospect isn't going to honor his or her signature, then all the contracts in town won't do you one bit of good.

I prefer to think that most of the people in business are honest, and when they sign a letter stating their intentions, they expect to abide by it. If for some reason a condition is not met and there's no deal, then all the parties part friends because it was in the original agreement.

In all my experience I have had only three letters of intent that did not go to full contract. On two of them the prospects very much wanted to build. The first couldn't secure financing; the second had zoning problems that everyone thought could be worked out, including me. The third was a prospect who decided to back out. In all honesty, I never should have even tried to sell and sign him up. I was uncomfortable after getting the letter of intent signed. There was nothing much I could put my finger on, just a lot of little things that didn't quite add up. I was relieved when the prospect backed out.

I don't think three busts are too bad when weighted against the many jobs I've successfully built. Letters of intent are similar to competition; you can spend so much time and energy worrying about them that the job doesn't get done. A letter of intent need not be a one-way street for the contractor; he or she can have some conditions incorporated. For instance, you include the condition that if the project is not built for whatever reason, the prospect has to pay a predetermined price for the plans. This will allow you to recoup what is usually the largest expense.

The letter of intent comes in all sizes and shapes. You can build it right into your proposal, and all the prospect has to do is sign on the dotted line. You can use a type of form letter in which you fill in the blanks; although this is handy, I personally don't care for it. I've found each project has its own particular conditions, and it's much easier to compose the letter to meet the situation.

After some exposure, you'll know what works best for you. One good rule of thumb: If the time starts to stretch out after you have presented your proposal and it's difficult to have a letter of intent signed, then watch out. Everytime this happens, the job falls through! The process should not drag. You may not get the job—it may go to a competitor; but don't let the selling drag out. Selling is a percentage game, and you have nothing working in your favor when it drags out. Drop it and work on something else.

Plans and specifications will be the first order of the day after you obtain your letter. I'm not going to go into how you get them; suffice it to say that you operate in the construction industry so it should be no problem to have the plans prepared. We are concerned about how the owner (X is no longer a prospect, but an owner) and the drawings fit together. First, there will probably be a time lag while the plans are being prepared.

I can only speak from personal experience about how long this lag takes, but it's usually 2 to 3 weeks, depending on the complexity of the project. I have had plans in as little as 1 week for a simple box warehouse.

One word of caution, though: Don't let it take more than 3 weeks to 1 month at the longest. I've noticed that the momentum starts to lag after about 4 weeks. This shouldn't affect the contract, but it could give the owner ideas about maybe delaying the actual construction for some reason. Don't forget you're dealing with businesspeople who won't see things in the same way as you do.

I use plans mostly to keep the enthusiasm alive during this phase. I always have the architect prepare a preliminary floor plan to 1/4-in scale. Sometimes I'll include an elevation that shows storefront details if I feel it's necessary. I have these drawn as quickly as possible, and I'm back in to see the owner in about 1 1/2 weeks. During that 10-day period I will have seen X at least once and maybe twice. There is always something to check on.

Be prepared for a very pleasant reception by the owner when you walk in with plans rolled up under your arm. This is the only time I've seen retail merchants tell someone else to take over and escort me to their private offices while customers were waiting. Nothing else will turn an owner on like the first viewing of the plans. I think the plans themselves are responsible for this attitude. When the owner finally sees the first tangible proof of the reality of the project, the owner is like a kid on Christmas morning.

Once I had an owner bend over the desk and push everything to one side when I presented the plans. Most of the paper and other items ended up on the floor. When I left an hour later, the owner was just sitting and looking at the blueprints. I hope you enjoy as much as I do this part of construction selling.

The preliminary plans are for the owner to check, who hopefully will make what few changes are needed. While this is being done by the owner, the site plan work is going ahead; and by the time the owner accepts the building plans, the architect should be ready to proceed.

I use the period of waiting for the final plans to have the owner mark up a floor plan for the phone company and have the owner confer with equipment suppliers if they have work to be done during the construction phase. For example, in the case of a retail tire facility with car lifts, preliminary work has to be done along with the slab. The owner should be involved with all such situations. It keeps the owner in the picture. She or he knows what's happening, and the momentum is steady.

When the final plans are completed, sit down with the owner and go over them in detail. Chances are, the owner won't know 50 percent of what he or she is looking at. Nevertheless, emphasize the areas that may cause problems and hope the owner understands. Many times owner X thinks he or she understands but doesn't have a clear picture. Those blueprints to you are crystal-clear, and construction people sometimes think everyone can read plans as well as they can. Most of the time X would not even know the bottom of the page if it weren't for the title block, so practice patience. You may find a method that I use helpful. When room sizes are discussed, I try to give the owner a visual aid. For example, if the new office is 12 ft by 14 ft, I look around to find some way to illustrate the size. I may state the dimension of the room we are using as a comparison. Size should be gone over with the first preliminary drawing and again with the final one, in hopes the owner has at least some idea of what she or he is getting.

Many times your best will not be good enough. You've done everything in your power to make X understand the showroom ceiling will be 9 ft high, but when X sees the grid being installed, X will tell you it's too low; X thought it was higher. You can refer to the plans as much as you want to, and X will always say the same thing: "I didn't understand 9 ft was so low; I can't read blueprints. You're the contractor; you should have told me 9 ft was so low." Be prepared for it, and if possible, put a little money in the job to cover this type of problem.

After going over the finished drawings, both you and the owner sign two sets of plans. You take one set and X takes the other. I always file my set and make a point not to let it out of my possession. This signed set of plans never gets to the job site; it stays in my office. If it ever happens that the owner tries to put one over on me, then I can haul out the plans and show X, or, more importantly, a third party, just what the owner had agreed to. I've had to do this only a couple of times, and in both cases everything blew over after the owner saw what he had agreed to.

I said earlier that you may want to have a condition in the letter of intent that reimburses you for the cost of the plans if the job doesn't go to contract. In special cases I can see where this might help make up the prospect's mind to sign. Maybe it's some complicated type of construction that's very difficult to price until the plans are drawn. Therefore, the prospect won't have a handle on price and has to take this route. The problem is you may find yourself providing a plans service, and there's no way to get back what it's worth for you to solve the initial problems that precede the letter of intent.

I really don't like any agreement where the prospect can buy the plans and go his or her merry way. I stay away from this and do it only when I think the job is worth the calculated risk. I'm the first to say this is another of those gray areas for the construction sales agent. The only guideline I can offer is to use common sense and know whom you're dealing with. If you do enough construction selling, you will be taken advantage of every once in a while. Expect it.

I personally know of some contractors who will prepare plans and tell the prospect to use them to take other prices. If their price is not low, then they are paid for the plans and the owner contracts with the other person. My answer is that I wouldn't do it under any conditions, period. You're right back on the bid market—and providing the plans as well.

The prospect can sometimes find more objections to signing than you can imagine. Most of the time this comes from a good case of being a little overcautious. You have to sign up prospect X while giving X an out. X will not be completely sure until X has the bottom-line contract price. Recognize this fact about the person you're dealing with, accept it as part of the program, and go ahead and work around it. When X knows the final contract price, X won't hesitate to commit completely. Don't try to force something that X is not comfortable with. If you do, you can blow the whole deal.

Contracts

There is one other way to sign up the prospect besides the letter of intent—I call it a simple letter contract. It's very practical for simpler jobs where you can nail down cost with your sketch. I use this mostly with metal buildings where there is very little subcontract work to be done—just slab, building, and erection. It's very easy to give the prospect a firm price under these conditions. Prospect X can see at once exactly what X is buying, and plans are not necessary to close the deal. X signs on the dotted line, and you're in business.

I've found from experience that the prospect is very comfortable with this type of special contract. X doesn't have to have a lawyer check over the fine print because there is none. It's clear-cut, and X knows where X stands. I even include a payment schedule so that all parties know when the money has to change hands. Don't try to cram too much into a letter contract. As soon as it loses its simplicity, its effectiveness is lost.

It is possible to sign up the prospect with several other contract variations that enable the project to get moving without a concrete price being determined. There can be dozens of reasons for using one of these methods.

The first is the age-old standby, cost plus. There's no need for me to explain how it works—you're in the construction business. If you can get a cost-plus contract, then good for you. It's an ideal way to do a design-build contract.

The other two types of sign-up contracts are actually similar. There's the contract not to exceed an established figure. I've used this type only a couple of times. The problem is that to protect yourself, you have to jack up the price. This higher price then may become an obstacle, so it requires a special situation. For example, once I used this type of contract because the prospect was under the gun with a deadline and dealt in perishable goods. For this reason she couldn't rent suitable space to move into on short notice; she had to have the new place. So in order to crank up the project at once, we agreed on this particular form of sign-up contract. I remember we went to work on the site with a line drawing of the building layout; by the time the plans were drawn and the permits issued, we had everything ready to pour the foundation. I might add that the city inspectors were very helpful and understanding. You may have a chance to use this type, but I'm sure you won't do it very often.

The next method is a variation of what I have just discussed: contract not to exceed established price, with savings shared. I confess I haven't done a project using this idea, but I do think this type of contract has merit. It should make the prospect comfortable to know that the contractor has an incentive to try to save money. The main purpose of this contract is speed. This is the only reason, in my opinion, a prospect will move ahead without knowing the firm price of the facility.

The whole idea in any case is to get the prospect to commit to you and allow you to put the project in motion. Use whatever method will obtain results. Familiarize yourself with all the various ways to sign up the prospect, and don't hesitate to use something new if you feel it will do the job. Each prospect is different; you are dealing with individuals in a one-on-one relationship. Approach each one with her or his requirements and problems in mind.

The prospect has been sold and signed. You've solved all the problems. The first price has been put together. It's now time to have the owner sign an ironclad contract so you can start work. What kind of contract do you have in mind? What kind of question is that, you ask? A contract is a contract. What's the big deal?

The big deal is that there is more than one type of contract. I'm really concerned about not the legalities of the contract, but how its form is accepted by the owner. From experience I know the simpler the contract, the more quickly it's signed. This is most important because you don't have a job until the contract is executed. The longer this takes and the more new parties are brought into the picture at this time, the greater the chance of delays and problems.

I imagine most of your contract experience has been with long, very detailed documents following the American Institute of Architects (AIA) or Associated General Contractors (AGC) format. The contracts may have been the type furnished to general contractors, or the contracts may have been prepared by the owner's lawyer in conjunction with the architect. Maybe your attorney drew up the papers. In any case, legal words prevail, and the contract is very likely to be difficult to read.

Another situation you may have encountered when using some of the standard contracts in the construction industry is having to delete whole sections that are not applicable to a particular situation. The owner can become uncomfortable when looking at a contract with sections marked out. Owner X is down to the wire now and wants to know that everything is exactly as it should be. The owner who isn't sure will turn to a lawyer and ask that the contract be checked over. The problem here is time; an attorney is not going to peruse the contract lightly. She or he will go over it with a fine-toothed comb to protect the client. This takes time, and everything stops until lawyer and client are satisfied with the contract.

If you are seriously going to pursue selling metal buildings and you do not use a shortened, easy-to-understand contract, I recommend you look into the matter. From experience I've found out it does help move the project along.

I personally prefer the letter contract. (Note sample on pages 184 and 185.) It's short and to the point and can be easily expanded to fit the job. A simple job consisting of a metal building on a slab for industrial storage takes up 1½ pages. On the other hand, a more complex project may take four or five pages detailing the work with specifications sheets attached. The important fact is that it all takes place in the letter contract form, which keeps things easy to understand. This means the selling process goes more smoothly.

As I pointed out, the letter contract on simpler, easy-to-price jobs can eliminate the need for a letter of intent. Added to this is an interesting fact concerning repeat work. Hardly ever will a letter of intent be needed when you are asked back to do work. The fact that you have established a very good relationship with X of which mutual trust is a large part will preclude the letter of intent. It's quite pleasant to get a call to stop by to work up the details, and then go from there to the contract. So my advice about repeat work is not to use the letter of intent except in very special cases. It's not usually needed when doing maintenance and repair work either. (See sample on page 186.)

There is one negative condition the owner always seems to get around to when the contract is to be signed: a penalty clause for late completion. It comes up so often I now bring the subject up for discussion before the owner does. I've never had to construct a design-build project where I worked under a penalty clause. I stress to the owner that the design-build job has the advantage of constant attention; if there are any delays, they will be due to matters beyond my control, such as weather or strikes. In addition, I always tell the owner I do construction for a living; if I don't perform as I say I will, then the job starts costing me money. So it's to my personal benefit that I finish the project as quickly as possible.

Being a businessman, X will usually accept these explanations; but every now and then you'll get an owner who'll bear down on the penalty clause.

LETTER CONTRACT

COMPANY LETTERHEAD

Date

Mr. X
ABC Distributing Company
Address

Re: Construction of new 6000 ft^2 metal building

This document is a contractual agreement between ABC Distributing Company and XYZ Construction Company to construct the above-noted building.

Details are as follows:

Metal Building

1. Furnish and erect one (brand-name) preengineered building 60 ft by 100 ft by 16 ft. Roof slope to be 1:12. Wall color tan, trim brown. Color samples included.
2. One (1) 3070 metal walk door with lockset.
3. Gutters and downspouts.
4. Two (2) metal overhead doors, manually operated 10 ft wide, 12 ft high.
5. 3-in white vinyl-backed insulation in roof and walls.
6. One (1) 3070 glass storefront door.
7. One (1) 6-ft by 6-ft and four (4) 2-ft by 5-ft insulated windows.
8. Three (3) future overhead door locations.
9. Eight (8) 3-in by 10-in skylights.
10. 4-ft roof overhang in front with soffit.

Concrete Work

11. Foundation will be steel-reinforced 3000# concrete; slab to be 5 in, 3000# concrete with 6-in by 6-in #10 wire with 6-mil polyvapor barrier; joints to be cut; all in accordance with plans by a registered architect.

Contract Price $90,000

12. Payment schedule $ 4,500 Signing contract
$42,000 7 days after building delivery

Monthly billing for all other work. Construction time approximately 90 days.

Items Not in Contract

13. No interior work in building (electrical, HVAC, plumbing, etc.).
14. Cost of permits, city fees, and site plan.

General Information

15. All proper insurance as well as builder's risk in force during construction period. Certificate to be supplied.
16. Plans by a licensed architect.
17. Standard construction industry 1-year warranty on all materials and labor.
18. Contractor will not be held responsible for any delays due to weather or areas beyond his or her control.
19. Building to meet all building code requirements.

ABC DISTRIBUTING COMPANY	XYZ CONSTRUCTION COMPANY
Owner X	Company Official

(Continued)

Here I use a tactic that's never failed. I say that since X insists on a penalty clause on completion date, and is a fair-minded businessperson, I'm sure X will agree to a bonus for early completion if I'm willing to agree to a penalty for late completion. No one has taken me up on my offer.

I'm not trying to tell you how to run your company business, but I am telling you ideas that I've found to be very helpful in closing the deal. I've talked with people who manage sales representatives, and they all tell me the same thing: One of the hardest thing to teach sales agents is the art of closing the deal. It's amazing the number of good salespeople who have trouble getting the prospect's name on the contract. I don't think the word *good* should be used in this case, for if they can't close, then they certainly

MAINTENANCE AND REPAIR WORK CONTRACT

BOOTH CONSTRUCTION LTD.
Specializing in Metal Buildings

This is a contract to replace the damaged wall sheets, trim, and gutter, located on the back wall of the main warehouse.

It will take 2 weeks to receive the replacement parts after they are ordered, which will be the same day this contract is signed. Work will start the day after the parts are delivered.

Insurance in place with XYZ Insurance Company.

Contract price	$4,200.00	
Payment schedule	$2,000.00	material delivered
	$2,200.00	job completed

____________________ ____________________

ABC Supply Company   Booth Construction Ltd.

aren't good sales agents. When sales managers discuss personnel, the bottom-line question is: Can the agent close? Only closers get the fame and fortune; all others should try some other occupation—there's no way they can be considered sales agents.

The worst mistake to make is to become complacent after you have the letter of intent. Keep right on pushing ahead because you have not closed the sale until a contract is executed. With this goal in mind, I prefer to use the simplest contract possible that will allow me to close the deal as quickly as possible.

I use several tactics when presenting the contract. One is that I take the approach that even when the owner obviously wants to get the project moving, she or he still has a tiny cloud of doubt. Keeping this in mind, I don't just hand the papers to X and wait for X to sign. As far as appearances are concerned, my main thrust is not getting the contract signed. I take the attitude that the contract is as good as signed. I bring up subjects that will require the owner's help or a decision very soon. The metal building color if it has not been chosen is a good start. I ask whether the owner can have the color by a certain time because I plan to order the building just as soon as I get back to my office. Notice I don't say just as soon as the contract is signed. The impression I want to make is that it is a foregone conclusion the contract will be signed. I am mainly interested in taking care of the loose ends.

Now while we are talking about the building color, let me tell you about something I do with every contract and why I learned to do it.

First the why. A few years ago a wholesale supplier wanted a new building to expand business. So it was a large building in a new industrial park that could be seen from the passing highway. The metal building supplier had a bright blue on the color chart, and the owner picked this so the building would stand out and be seen by all passing by.

So the building was delivered to the job site, and erection started. A few days went by, and then the blue sidewalls started to go up. The owner and his wife came by everyday to check our progress. We had about 30 ft of blue sidewall sheets up when they came by and started looking. Then it happened. The owner's wife did not like the blue color. The owner told the crew to stop and have someone from the company call him. I will keep the story short. His wife said the sheets did not match the color sample we had used, and she wanted something done to make it look better.

Now the sample and the actual wall color were close, but not an exact match. However, color charts most of the time can be slightly off, and this is accepted in the construction business.

It took 3 weeks to resolve this situation and get the job back underway. Our problem from the contractor's side was that there was no written proof signed by the owner that he had picked the blue from the color chart. We actually talked about changing the color since not many sheets were up, but the owner saw the delay this would cause. He was under a tight move-in schedule, so he convinced his wife the color was OK, and we went back to work 3 weeks behind.

When this problem came up, we could have taken care of it in a few hours by showing them the color noted on the contract, either at the beginning or added later, even though the building had already been ordered—plus a color chart with the chosen color noted and signed by the owner and the contractor. With that proof the contractor is in control, and that's what is important.

Because of that episode, I now include with the contract a signed color chart that becomes part of the contract. So, note the color in the contract specifications but include the signed chart so there is no guessing as to what color is being used. I do this with every job, and since I started, I have had no other color problems.

Another good topic to bring up for discussion is the coordination between the contractor and people installing special equipment who are working directly for the owner. A good example would be a shop you plan to build that will have a bridge crane. The owner is dealing directly on the crane, but you have to provide special supports for the rails. Then when all else fails, and I really have nothing to ask the owner, I fall back on inflation. I tell the owner that the timing is perfect; with what inflation is doing to construction costs, the owner would have to pay a lot more for the facility had he or she decided to hold off. I use this only when I have nothing else to bring up, although I've found the owner will sometimes mention it while we are talking about details.

Don't be bashful about asking for the contract. A good closer will ask for the signature at the right time. It's easy to work it right in while going over details. I say something like, "X, I brought only one copy for you to keep. If you need more, I'll drop them off tomorrow." (Notice I did not say I would fax it.) While talking about the contract, I suggest that we go ahead and take care of the paperwork while we are on the subject. I learned from another sales agent that it's a good idea not to have signed yourself; doing so at the right moment will start the ball rolling, so to speak. I've used this suggestion successfully a number of times. Consider it.

Some of you may think all this to-do about how to handle the contract signing is a lot of nonsense. My answer is the same thing I've repeated so many times before: Selling construction is strictly percentages. You're always trying to have the percentages work for you. Consider this: In 90 percent of your contract closings, you are probably right; it's not necessary to go into such elaborate selling maneuvers. But for the other 10 percent, it very well could make the difference. The hitch is: Will you be able to tell which owner is in the 90 percent category and which is in the 10 percent? I can't! So I assume everyone is in the 10 percent group. I believe my approach will never cost you a contract. It can do nothing but help.

No matter what form of contract you use, they all say the same thing: what you will be paid for performing a certain job. And this leads us to another item that should accompany the contract: payment schedule. The payment schedule is important in a design-build project because the general contractor has already spent money to arrive at the contract stage. Also, there will be anticipated expenditures to start up and staff the job site. The metal building may require a deposit to be sent to the manufacturer prior to fabrication. To cover these costs, I always ask for and get a deposit when the contract is signed or very soon afterward.

Don't just spring this deposit on the owner. After the letter of intent and before the contract is ready, I discuss how the money is to be handled and exactly what the owner can expect. My rule of thumb for the deposit is 5 percent of the contract amount. This 5 percent figure is flexible, though; sometimes I'll decrease it, depending on the project and the owner.

I take pains to explain to the owner exactly what the deposit is for—to pay the design professionals and to pay my company for time and work done to date. The owner is a businessperson, and when you explain things in terms she or he can relate to, there's no problem. The owner is in business for the same reason you are, to make money.

Retainage and Bonds

I know you are familiar with retainages. They're something we have to put up with in the construction industry. You'll be faced with retainages in design-build as well as bid work. It's brought up by the owner, and if the owner insists, I put a 10 percent retainage clause to be applied to each invoice. I only do this now when the owner prefers it; usually when the subject comes up, I try to persuade the owner that it's not necessary on a design-build contract because he or she knows the parties. It is really used for owner protection on the bid market, where the contractor is probably a complete stranger to the owner. Frankly, most of the time the owner accepts this explanation, and there is no retainage.

However, there is one thing I make clear when I do have to work with retainage being withheld. When the job is complete and the owner moves in, the full amount is due. No retainage will be owing when X enjoys the use of the new facility. Callbacks will be handled on a warranty basis, not by holding my money as a club. Also, the question of a bond may come up, if needed. Then I tell the owner the cost of the bond will be added to the contract as an extra.

Changes

This is a good place to discuss a peculiar situation that you'll run into if you do enough construction selling: After the contract is signed and the job is coming out of the ground, the owner thinks he or she can start asking for little favors that won't cost anything. Maybe you've already encountered this type of owner. In defense of some owners, they really have no idea what it costs to do something and think a favor or a small change won't make any difference. There's only a few bucks involved. Then you have the sharpie who's going to try to squeeze something out of you for nothing.

In the first case, if I can do what the owner asks and it involves only a few dollars, I do it. But if it's going to run into some money, I make it a point to explain, diplomatically I hope, that the change will cost some dollar amount. Then I explain why. It could be a very small material cost and high labor, or any number of variations; but the point is that I explain to the owner my situation.

I deal with the squeezer in a more blunt fashion: I put everything in relation to the business. If the owner is a car dealer, I ask how many extras are given away when a car is sold. In other words, ask the owner what she or he gives away, and the owner gets the picture very quickly.

Checking Owner Financing

Most of the owners will have a construction loan, so it becomes necessary to provide invoices at the end of the month for the owner to turn into the bank before making a draw. Therefore, I make a point of explaining exactly what X can expect at the end of the month. Check with the owner's bank before the first draw is due so that you will submit an invoice including everything the bank needs. You certainly don't want any delays in being paid!

The payment schedule can be separate and attached to the contract, or it can be incorporated into the contract itself. The letter contract can have the payment schedule included as well. (See example on page 166.) It's important that all parties know how the money will be handled. Always ask for and get your front-end deposit. It's a comforting feeling to be operating on the owner's money right away.

I generally put up with some headaches from the owner if I know the money is there and waiting. The way I do this is to ask for a copy of the mortgage commitment and confirmation from the bank making the construction loan that everything is in order. I like to do so diplomatically because I'm faced with the same problem I have when presenting the contract to the owner: 90 percent are going to sign, and the remaining 10 percent may have some reservations.

The problem with getting into the owner's finances is that some businesspeople are touchy about their word and their credit. Every now and then you'll encounter someone who becomes annoyed when you doubt his or her word by asking for proof. Most of the owners accept providing proof as the way business is done, especially when dealing in the sums of money involved. Again the problem is who's touchy and who's not. This becomes important because the situation usually arises before the contract is signed.

I've developed a tactic that works: I become an agent for a third party who needs the information. The first thing I do is to take the offensive. I point out that I have to have a copy of the mortgage commitment. Then I explain that my bonding company would like to have the information, or I might use my suppliers. I explain that because of the large amounts of material I'll be buying from them, they want to know the bills can be paid if I, the contractor, drop out of the picture for any reason.

Now X may get a little annoyed, but not at you—rather at a third party who is just a name. You come across as a nice person trying to do your job, and these other people have put you in the middle. X does as asked more to help you than for any other reason.

This chapter is for the benefit and protection of the contractor. A signed contract is the target you were shooting for when you started chasing down leads. It's the bottom line, so use my suggestions if you feel they will help. Along with the contract, make sure the owner knows how you expect to be paid, and protect yourself by knowing the owner has the money to pay you. Assume nothing; no one else will look after your interests as you must yourself.

Let me sum up this chapter on the paperwork by stressing the fact that I am sure you are familiar, and have plenty of experience, with construction forms and contracts. This means you will have no trouble coming up with what will do the job to get the project signed up.

The main thrust of this chapter is to combine selling with signing up. I strongly feel it is just as important to be selling you and your company here as in any other area of design-build marketing.

So regardless of what particular form of paperwork you choose to get the project moving, I suggest you keep the selling recommendations in mind.

NOTES

NOTES

11 Building-Phase Communication

With the signing of the contract, you've accomplished what you set out to do—sell a design-build project. The active aggressive selling is ended; all that's left is actually to build the facility. Right now if you're like most construction companies, you file away the contract, assign a superintendent, and move on to something else. And if you do, I'll guarantee that at some point during the job you'll have trouble with the owner.

Contractors have a difficult time realizing the owner in a design-build project has to have his or her hand held from start to finish—I mean right up to the grand opening. I think maybe the standard attitude of contractors comes from the manner in which they have conducted their bid business. The owner is really a stranger kept at arm's length with the plans, specifications, and contract governing every move. It's natural for contractors to operate in this way; it's the way it's always been done. From a sales standpoint, however, it's absolutely the worst way to handle a design-build owner.

On the other hand, the owner of the construction company has problems, too. The owner can't drop everything to look after one customer who might get his or her feelings hurt. Granted, he or she can't stop everything, but it's still possible to give a little extra attention if the effort is made.

Start-up

Within a week after the contract signing and preferably within just a couple of days, there should be some sign of activity on the site. The best thing to use is a job sign, and a trailer if one is required for the project. There can be real trouble if nothing seems to happen on the site; the owner will get upset, which many times will set the tone for the entire job. It's important for the owner to see attention given to the job.

During this period, check whether the owner plans a ground-breaking ceremony. If so, then offer your help, of course; and, more importantly, work out the timing so the job signs are up and the top person in your construction company is there to get his or her picture in the paper. This is great free publicity; take full advantage of the opportunity. Don't hesitate to bring up the topic with the owner. The owner just might decide to go ahead with your help, and the pictures in the paper will be money in your pocket.

Along with these visible signs that the project has started, make every effort to get the project out of the ground. Any good construction person knows it's advantageous to get going, but things do have a tendency to drag sometimes. If the problem is workforce, then make sure the design-build job takes preference over a bid job, if possible. Treating the bid owner in the regular manner won't change things, but treating the design-build owner in the regular way could cause problems.

Work Schedule Information

This project is the biggest thing in the owner's life right now. X expects to be kept up to date on progress. Often that's vitally important to owner X because X has to coordinate the moving of the business; the buying of new stock, the delivery of new equipment, and even being able to terminate a lease all make it imperative that the owner have a good idea of the construction schedule.

I've learned that my presenting the owner with a truthful schedule is something very much needed and appreciated.

Acquainting People on Site with the Owner

There's no way you're going to keep the owner away from the site during construction. Don't even try. Instead prepare for her or his being there. Always let your superintendent or foreperson know, if possible, that you and the owner or the owner alone will be coming to the site. Make sure you introduce the owner to your key people on the site. Make sure the construction people understand X is the owner and that when they see X on the job, they are to make it a point to speak and try to be helpful. (See Fig. 21.) Under no condition is the owner to be ignored. It's the responsibility of the person in charge to take care of the owner when he or she arrives.

Let me digress for a moment with some friendly advice for the construction salesperson who will be working for a construction company.

Figure 21. The owner and family watching a building construction.

First realize that the bulk of the hired hands will feel just a little bit of resentment toward the sales agent. There is no way you can convince them you work just as hard as they do. They see you as a good-time Joe riding around with owners, showing them jobs and then maybe playing golf on nice afternoons. You don't get your hands dirty or freeze in the winter and bake in the summer. These workers know beyond a shadow of a doubt that you make as much money as the owner of the company.

With the workers having this opinion of you, take it nice and easy when you visit the site with or without the owner. Treat everyone as an equal; get to know the foreperson if you don't already. Don't tell the construction people how to do their jobs or act as a spy for the owner of the construction company. If something is obviously wrong, question it; but don't go running to the office. Let the foreperson handle it; that's her or his job. I always made it a point to ask the person in charge on the job site if there are any questions I can help with. Invariably on a design-build project, changes have been made that create questions for the people driving the nails. In like manner, when a change is made during construction, I visit the site and, over coffee, update the superintendent. This makes his or her job easier and cuts down on costly mistakes. At the same time it shows the nuts-and-bolts crew that you're actually involved and are working just as hard as they are at keeping the job running smoothly.

Acquainting the Owner with Construction Problems

Explain to the owner that not everything is going to be rosy all the time. It's important that the owner be made aware of the basic problems that always seem to slow down a project. The only yardstick the owner has is her or his own business, and chances are that it is an inside operation. The weather does not shut down the owner, and when there is a problem with goods, the owner takes another item off the shelf. The owner is used to solving business problems by taking immediate action.

I explain to X by relating my problems to X's business. I ask what would happen if an inspector came in and, because of some fine point in the interpretation of the code for selling tires, made X shut the doors until the matter was corrected. How would X like the idea that every time it rained X had to stop business, the production came to a halt, but X had to continue to pay some of the help? After a few minutes of this, the owner starts to have a better appreciation of what the contractor faces. It really is another world!

During the construction phase, keep the owner up to date on the bad news if it affects the completion date. Don't let X find out from someone on the site that the metal building will be 2 weeks late. You'd better know yourself and then inform X, and just as important, tell X why.

At times a picture will be worth a thousand words. Take the owner to the site and show X the situation. I remember a few years ago it seemed to rain every day for 3 months. We were getting further and further behind and there was no way to catch up and make the projected completion date on a particular job. The owner was getting upset with our lack of progress. He simply could not see why, when it stopped raining, we couldn't get anything done.

I finally decided to show him exactly what we were faced with. I took him to the site where the half-finished building stood surrounded by a sea of mud. Sitting in the mud with the bottom half of the tracks out of sight was a bulldozer. I told the owner that we had tried to use the bulldozer to reach the building that morning, but he could see how far we went. I didn't hear another word from him about our slow progress.

Keeping the Owner in the Picture

As stated earlier, the owner must not get the impression of suddenly no longer being important. It'll be up to you to keep X very much in the picture.

At the end of the month when the invoice is prepared, don't just mail or fax it to X. Instead, call and ask if you can pick up X for a job visit. Tell him you have the monthly invoice, and you want to check it out with X. Now the invoice should list the work that has been completed or percentage of completion if it is still in progress. The idea is to show the owner what the invoice covers. This gives X a good indication of just what progress is being made. Chances are, the invoice will end up at the bank, and the bank people will inspect the site for themselves. Nevertheless, your customer is the owner, not the bank, so cater to the owner.

As the job progresses, small questions will arise that may or may not require a decision by the owner. For example, the swing of an office door is in question because a vent had to be moved; or there is a question about where a vision glass should be located in the shipping supervisor's office. These are the little questions that put the owner on center stage and keep her or him feeling very much part of the project. Even if you already know the answer, file the question away and use it when it's time to involve the owner.

Whenever possible, have the owner visit the site to make the decision, but if that is not always convenient, go to X's place of business with your plans. The owner should see you at least once a week. Try not to make up trivial little things to present to the owner. Use only legitimate questions, as the true professional does.

When it's time for interior selections to be made, such as paneling for the walls, paint colors, and floor coverings, you have a very good reason to consult the owner. Again, don't just drop off the samples. Go over the choices, and if you know where a certain item has been used, offer to show it to X. Stay involved and be helpful at this stage. One word of advice, though: Do this well in advance of when it is needed, or you may be in a pinch for time, and the owner could end up slowing down the job until final selections are made.

As a matter of fact, you might use the special inspection as a reason to get the owner to the job, especially since X could be paying directly for the service. This will show X that X really is dealing with professionally oriented construction people looking after X's interests.

Don't worry too much about what you'll use for reasons to call on the owner and keep her or him in the picture. With a design-build project, it's amazing just how many situations arise that have to be checked out. The point to remember is that the owner has to be kept in the picture. (See Fig. 22.)

Owner Problems during Construction

As much as you and I would like to think that this subject is unnecessary, unfortunately that just isn't so! Take note, and believe me, you will have

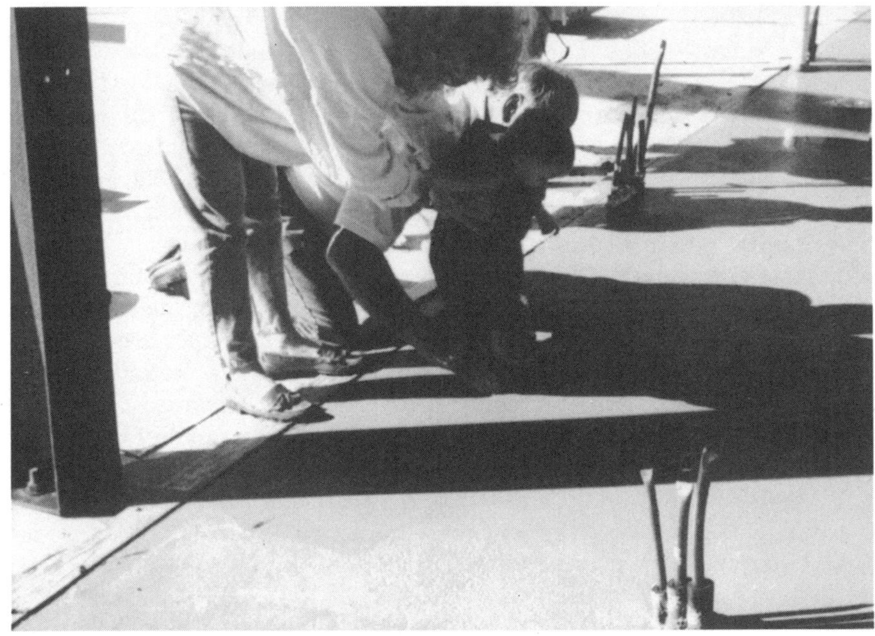

Figure 22. Family members put footprints in the office slab.

Figure 22. *(Continued)*

problems with the owner. The very best you can do is to anticipate them and be prepared; hopefully staying on top of the job will allow you to catch a little problem before it blossoms into a monster. Below are the major problems with the owner that I have encountered, along with strategies to handle the situations.

The Plans Don't Agree with the Owner

This problem always seems to come up, and late in the job when X can actually see exactly what X is getting. The basic problem is the inability of the owner to read blueprints. It's nobody's fault; it comes with the territory, and the contractor has to accept it.

I touched on this problem earlier in the book, but I feel greater detail is needed. The first defense you can use is to keep drumming into X's head what the plans call for. The second and third defenses are what I mainly want to cover. The second defense is to be astute enough when going over the plans with the owner to recognize when X is really not getting the drift of what you're explaining. The owner might say he or she does, but you know differently. This takes a little experience, but it does come with doing design-build jobs. When you sense this, or know it's a complicated area you're discussing. write it down to follow up. When the area in question is being built and the shape and size can be easily seen, bring in X and go over the section so that X can see and understand what she or he is getting. The trick is to do this soon enough that if changes must be made, it won't cost too much.

When the question does pop up, you have to fall back on the plans. They show exactly what you priced and will be the basis for a change order. The owner may grumble some about assuming this and thinking that, but will be able to see your position.

Every now and then an owner won't be satisfied with the way the plans read. X will keep coming back to a difficult-to-answer statement: "That's not what I told you I wanted; never mind pointing to the plans. I know what I asked for." What do you do? Fall back on the plans and prepare for a fight? Not if you can possibly avoid it. This leads to our last defense—the slush fund, which is a small amount you included in the job to cover such an unpleasant situation if, when down to the wire, you felt you had to give. Don't give until you've tried every avenue of compromise, such as offering to make the change for cost or maybe splitting the cost. In any event, for the sake of good relations, use the slush money. Cheer up; maybe the owner will ask for a legitimate change order, and you can get your money back.

A great deal of the way this unhappy situation is handled is up to you and depends on how well you know the owner. I haven't had many owners who really presented a problem, and every time I did what they wanted, it was simply for public relations. Sometimes I gave more than I should have, but I was not forced to. There is a difference.

Change Orders

I know you're familiar with how to prepare change orders: in writing, with everything spelled out and signed. But I would like to mention *tolerance.* Sometimes it's easy to become upset with the owner over a change order. It's a pain in the neck for job progress to be interrupted by a change order. Nevertheless, the design-build owner deserves and should receive cheerful handling of change orders. Show the same consideration you gave the owner before the contract was signed. It will all add up to recommendations in the future. Never forget how important this is.

The Construction Schedule Falls Behind

Despite the general contractor's best intentions, the majority of projects seem to fall behind. I firmly think the reason is that time is used as a selling tool. I know I've done it! You don't want the prospect to become discouraged over the excessive time span, so you present the best possible construction time frame and hope you don't stumble. Well, somehow you do stumble, and the job starts running behind. The owner, with all the commitments, suddenly sees a crunch coming down the road and starts letting you know about her or his problems.

I know you've been in this box before, but let me tell you how I handle it. Even when the job is coming out of the ground, I tell the owner every time there's a problem beyond my control that costs the job time. Then X is aware and can judge the commitments made for X's openings.

The weather, more often than not, is the main reason for the job slowdown. I use what I consider the ideal way to handle this with the owner. The idea comes from a company I worked for. The weather report was clipped from the paper and filed every day. At the end of the month a tally was made of the days it rained and the clear days. The strength of the wind was also noted, since it has a direct effect on metal building erection. These weather notes will give you good backup when you meet with the owner, so take the time to clip the information.

Now the owner has no trouble understanding bad weather as the reason, with one exception. Just as soon as the rain stops, X expects the job site to be covered with workers. One of the hardest points to get across to X is that the side effects of the rain are just as damaging as the rain itself. So when you explain to the owner some of your problems, stress the side effects of rain and how they can cripple a job. Even then you get phone calls from X demanding to know why there's no work being done, since, after all, it stopped raining 20 minutes ago. My advice is—be nice!

On this subject of falling behind, I've made the assumption that the delays are beyond the contractor's control. If they are within your control and the job is behind because you're not pushing it as you should, then you'll have a real problem when and if the owner finds out. This is the one area where the entire relationship you have so painstakingly built up with X can turn into a can of worms. From this point on, you're just like any other contractor, and the owner feels justified in doing anything possible to protect her- or himself. Guess what X uses for protection? If you said money, you're correct. X starts dragging his or her feet on payments just as the job is dragging.

Finishing the Job and Punch List Performance

In my opinion the last 5 percent is always the most difficult to get completed. The project is winding down, and hopefully you are involved with new jobs coming out of the ground. Therefore, X's job is receiving less and less of your attention.

It's extremely important for the contractor to work at staying on top of the finishing-up process. I know personally how hard it is when you have a full load, but it has to be done. I learned the hard way by having a couple of jobs turn sour in the last 5 percent stage.

Not only do you stay with the project, but also you give a little more time to the owner. It's a very trying time for X, who is facing the hassle of moving the business, plus having to coordinate with suppliers. Some personal attention and help with getting the building ready will be appreciated when referral times come round.

When the project is very near completion, arrange for a walk-through meeting with the owner and foreperson responsible for seeing that the work is completed. Give yourself plenty of time. These punch list walk-throughs can run into time because the owner will be enjoying the new building. After the punch list is worked up, get all the work done in a reasonable time frame. Arrange a second meeting to go over the completed punch list. Check the job ahead of time so there won't be any surprises.

Finally X is in the new building, but the contractor's work is not done yet. Two other matters should be attended to. The first is the owner's

Figure 23. The owner's grand opening.

grand opening (Fig. 23). If there is one, be there. X is pleased as can be with the new facility and wants very much to show it off to all comers. As close as the owner has worked with the contractor during the job, X will feel slighted if the contractor doesn't take the time to attend and help celebrate the occasion. There's another reason to show up: It's a great place to pick up leads, so mingle with the crowd! There's a good chance some of X's competitors will stop by to see the new facility, enabling you to do a little self-serving lead gathering. So have plenty of calling cards in your pocket to pass out.

The second chore left for the contractor is to follow up in about 3 months with the owner to check on any problems. Most owners have never known of a contractor to check back. This follow-up call really impresses the owner. I've picked up several good referral leads from an owner while making this type of call. You may wonder why the owner didn't call you about the lead, but probably the owner just forgot it until he or she saw you.

Also there will be callbacks right after the owner occupies the building. See that they are taken care of, and then personally check back. It's this kind of service that reputations are built on; it means business in the future.

The contractor who sees the owner through all the trials and tribulations of construction will be making sure that owner becomes a ready source of business. The happy, smiling owner will be your best sales representative. This happy result doesn't come free; you have to work for it, but the rewards are well worth it.

NOTES

NOTES

12

Metal Building Sales Department

This chapter is intended primarily for the construction executive who is interested in forming a sales department. When I mention sales department, it can mean one or more sales agents. If you have or are planning to employ one salesperson, and the responsibilities will be selling, then you have a sales department.

I have no doubt whatsoever that more and more general contractors will be using professional sales agents in the future. As time passes, we will experience many changes throughout the industry, and construction selling will be a part of these changes. I hope my thoughts and the experiences that I relate in this chapter will help you form and operate a successful sales department.

The first thing I should describe is the type of person you'll employ as a sales representative. You'll have someone working for you who will look as if she or he is not working. The agent will wander in and out of the office on no set schedule. Most of the time you'll have a vague idea about where the agent is and what she or he is doing. Sometimes you won't have the foggiest notion of where the agent is. You may go for a couple of days without seeing the person. You call the agent's home in the middle of the morning because you can't get through to her on her mobile phone, and the agent answers the phone. All this can be difficult to accept if you are used to people punching a timeclock. But stop and think for a moment. Where is the one place your sales representative is not going to locate business? Your office! The office is something the agent will use to help him- or herself. It becomes only a backup operation to the main working area, which is everywhere except the office.

I'll never forget what one of the owners of the firm where I first worked said to me when I started. He told me the one place he'd better not see too much of me was in the office. Neither he nor anyone else in the office had any intention of buying a building from me.

The truly professional salesperson is very independent and wants very little to no supervision. This is even applicable to firms that use a formal sales force, such as the building material manufacturers and metal building companies. They have sales managers and all the necessary chains of command, but their sales manager knows when it is a professional working for her or him. It isn't necessary to read the paperwork to see what the person has been doing; simply glance at the order sheet, because that's the only yardstick to use to measure a professional sales agent. Does the person get results? That's the bottom line; everything else is just window dressing.

A good one-word description of the construction salesperson is a *maverick.* This agent is her or his own person, a self-starter who's good at the job and knows it. The construction sales agent is told what is expected to be accomplished but not how to do it; that's up to the agent.

Okay, you say, I get a picture of what I can expect of this individual; but how do I fold her or him into construction company operation, and, just as importantly, where do I find such a person? The rest of this chapter will help with these questions.

Integrating the Sales Department into the Overall Operation

Your first step entails the physical part of integrating a sales department into your company. The sales agent needs a place to take care of paperwork and make phone calls. An office is fine, but if one is not handy and you have a conference room with a phone, there's nothing wrong with using it. As a matter of fact, I used a conference room for 2 years as my office, when I was a metal building construction sales representative. When the conference room was in use, I took over an empty office. This worked very well, and the firm didn't have to rent more space, which helped on the overhead.

After giving your salesperson a place to hang a hat, see that the clerical requirements are covered so that necessary office work will be done. With regard to typing, it's important to give the sales agent some rank. The agent needs to have some authority to be able to get office help on short notice. And I mean short notice! When the prospect tells your sales agent to type up a letter of intent to sign, it had better not end up on the bottom of the typing pile. It goes right to the top and is finished promptly.

The sales agent has to be able to have a short turnaround time in the office when material is needed to get a job moving. Many times it's simply a case of asking the typist to please do the sales proposal next. In larger offices where there is an office manager, let the sales agent go through the manager. You have to make clear to the manager, though, that the sales request will be treated as priority when he or she is asked that it be handled in that manner. Make sure your sales agent understands not to ask for priority treatment unless it's important. Sales representatives have a tendency to be impatient when it comes to paperwork, expecting everything to be done for them immediately. There will have to be a little give and take here.

The office help won't understand why this new person has to have a letter typed in the next hour. They are used to having the paperwork move at a set pace. The people doing the work and the people receiving the material all know how the system works. Nothing, they will think, has to be done that quickly; after all, the salesperson has been working on that particular prospect for the past 4 months, and a couple of days certainly won't matter. Well, it just might matter a great deal, especially if it's toward the end of the week. You will have to impress on the office people just how important some of the paperwork will be.

Even then, be prepared to referee every so often; for a good salesperson will take matters into her or his own hands when necessary. For example, I had been working on one prospect for 6 months, and I was determined to sell the job. One Friday morning X told me to work up a letter of intent and general specifications to go with the drawing I had supplied. I asked if X would be available that afternoon if I could get the papers typed up for signature. X replied yes, but what was the hurry. I said I wanted to work on the job over the weekend, which was only partially correct. I also wanted the papers signed before the weekend.

I drove directly back to the office, dashed in, and found there was a problem. It was the end of the month, and they all were working on invoices to send out. The office manager was a man who was very sympathetic to the plight of the salespeople (there were two of us) except when preparing month-end invoices. Then he guarded the typists' time fiercely. There I was with a hot prospect ready to sign, and no one to tell the typists to type for me. And their instructions were implicit: Type invoices and nothing else.

I took the attitude that the invoices of sold jobs should take a back seat to one that needed selling. I told two of the typists to stop what they were doing, and I put one to work on the letter of intent (the draft of which I had been carrying around in my briefcase for a month) and the other to typing a specification list. Just as they were finishing the typing, in walked the office manager. He took one look around and did a marvelous job of holding his temper. I snatched my completed papers, thanked the typists, and bailed out. One hour later I was back with everything signed.

I also knew I would be hearing from the office manager. Sure enough, I'd no sooner walked out of the boss's office that afternoon than he caught me in the hall. I listened politely while the office manager told me that he was going to have to pay overtime to the typists to make up for the time I cost him. He said a few other words which I'll not burden you with. When it was all off his chest, I asked him one question: "What would you have done in my place?" He grinned and said, "The same thing you did!"

The point I want to make is that the sales agent, if doing her or his job, will be a disruptive force from time to time.

The next step in bringing this new person into your organization is to explain to all on the nuts-and-bolts side how the salesperson fits in. Your superintendents on the jobs first will take the attitude that the salesperson will be looking over their shoulders and reporting back to the boss. The best way to dispel this notion is for the salesperson to establish a rapport with the field people. The main reason to check the jobs is to gather information to supply to the owner. And let the person running the job know this from both you and the sales agent. In the end it'll be up to the salesperson to integrate into the construction side of the company; and she or he should do it with ease, since the job involves dealing with people.

At the same time, if something is discovered that is going to affect the job and create problems with the owner, you'll be the first to know. If the superintendent gets his or her nose out of joint, then too bad—because your salesperson isn't going to stand by and let the owner get the wrong impression about the company. And if your sales agent doesn't take this attitude, then you don't want him or her. Aggressive selling means being aggressive in more than one area.

There is a negative aspect to adding a sales department. Hopefully the salespeople will be able to make it disappear, but I personally think it never goes away altogether. Every sales force is resented to some degree by the rank and file. By management, no, because managers recognize the contribution a sales force makes; but the people sitting behind the desk see salespeople as freewheeling good-time players taking a customer to lunch, then writing up an order with commissions amounting to what they make in a couple of months. Then these same people have to take care of the details to make sure the customer receives what was purchased.

This resentment can really exist in a construction company with the people working on the site. It can begin with the person in charge and work right on down to the worker with a hammer. Therefore, it's imperative to impress on your people in charge of the field just how important the salesperson's job is. Stress the concept of teamwork, that somebody has to build what the salesperson sells and that everyone is doing a vitally needed job. As I said, I don't think the resentment will ever go away completely. What you're working toward is to make it an insignificant factor and to obtain money-making results from your employees.

It's going to be up to you, the employer, to put the salesperson in the picture. Afterward it'll be her or his job to make it work. The sales agent should report only to you or top management; no one else should have one thing to say about her or his job. The sales agent's position is comparable to that of a staff job where one has the authority to get work done but has no one answering directly to him or her.

A Salesperson's Responsibility to the Company

As much of an individual as the sales representative will be, and as much as he or she will dislike paperwork, there will be certain duties that you should expect. Proper records are a must. It's vital for you to know what's being worked on and what the near future holds for contracts. The sales agent should keep lead and prospect lists with all pertinent information for each listing, status of the prospect at the last meeting, and what is to be done before the next meeting. The time of the next meeting should be recorded also. This database will serve two purposes: It will give you, the employer, a handy printout to check to get a feel for what the salesperson is pursuing, and it will provide the salesperson with a printout to use to keep all the loose ends under control.

Along with the prospect data it's very important to spend some time every 1 to 2 weeks with your sales agent and go over the prospects as well as owner relations with jobs under construction. This exchange of information can be very helpful to both of you, and all concerned will stay up to date.

Don't try to tell the sales agent how to go about selling. I've seen it happen. It's hard for employers not to express their ideas on how the salesperson should be doing the job. Convey overall expectations, but not the details. That's what you're paying the sales agent to know. So fight the urge to do it.

Communications during the day are pretty much up to the salesperson. I've found it best that he or she check with the office at least three times and never less than twice. It will be a good idea to tell her or him to always take the time to call the office. Now let me say here that any sales agent who is a pro will check with the office, since the office will be a source of leads. Of course, a pocket beeper should prove very handy for all concerned, as well as the mobile phone.

While working, the construction salesperson must be on guard against giving orders to the construction people. The pro knows how to handle the construction side of the business; but if you have someone coming on board who doesn't know very much about how a construction outfit operates, then explain what *not* to do. I know how difficult it sometimes is to keep quiet. Nevertheless, the sales agent seeing something out of line should go through the person in charge. Although the employees don't work for him or her, the salesperson is recognized as management; therefore, more likely the workers will accept orders from her or him. Thus, the sales agent will be undermining the superintendent, which is guaranteed to create friction between the sales and construction departments.

Sometimes people in the company will look on the salesperson as an errand runner. It's only natural to ask so-and-so to drop it off. This can get completely out of hand. It is best for all concerned to make it clear that the sales representative is not a messenger. Now I'm the first to admit there will be occasional exceptions that are not inconvenient or important enough to warrant it. What you must be on guard against is letting these exceptions become more frequent. You can tell the office people, and the salesperson as well can say no to the office staff; but how about you? You are the one person the sales agent works for, and he or she cannot refuse you; so it's mainly to you that I'm directing this advice. That sales agent is worth too much to be used in this manner.

Do not expect the sales person to keep regular office hours. He or she must be free to move where the business goes on the time schedule that gets the job done. As I said before, nothing is sold in the office.

Finding the Construction Salesperson

You're wondering just how you locate a construction sales representative. You know the local employment agencies won't be able to help you. This is simply too specialized a field. So you're back to square one.

It's not really that difficult if you know where and what to look for. Check around with the competition and determine who's using salespeople. If there are some operating in your area, you can approach them to see if any will consider working for you. This is probably the fastest way to hire someone. It can also be the most costly financially. Don't expect to bring in an experienced person for peanuts. The real problem is that there are not very many truly professional construction sales agents around, so chances are you'll have to create one.

If you list the people working for you, to think about who you can promote to sales, you may be wasting your time. You might get lucky and have someone on your payroll who can do the job, but it's a million-to-one shot. People with sales ability don't go to work for construction firms; nuts-and-bolts people work for contractors. Sales-minded people go where salespeople are used; so that's where you have to look.

The very best place to start your hunting expedition is within the construction industry itself as well as closely allied fields. I really don't recommend going outside this area because then the salespeople will know absolutely nothing about the construction industry.

You're looking for someone who's first and foremost a salesperson; that's the most important fact. Next you want this person to have some knowledge of construction. She or he doesn't have to be an ex-superintendent, but it helps if the agent knows the difference between a concrete slab and a built-up roof. You can teach a good salesperson the necessary construction details.

This salesperson must be used to selling to the person in charge. An order taker who calls on purchasing agents is not for you. Your sales agent must understand why she or he has to deal with the person who signs the checks. Next it's very helpful if her or his background has been in larger-amount orders.

You want a person who's been in your area long enough to get established. So much of construction selling is whom you know, and a newcomer will be at a decided disadvantage. A pro will pick up details of the job quickly, but if you have to train someone to sell, then you're wasting your time.

Age is a factor that will affect sales. I have a stockbroker friend who was delighted when he started getting gray at the temples. He also tells the story about another broker who had a little gray added over his ears to look older. Whether the story is true or not, the point is well taken. There are several selling positions in which maturity is a definite advantage, and construction sales is right at the top of the list. By far, owners in the market for a new facility will be around 40 years old or older. These owners will be graduates of the business school of hard knocks. Confidence is a large slice of the selling pie, and these owners, when talking about thousands of dollars, prefer to talk with mature people.

The whole thing boils down to this: You're looking for a mature, established professional salesperson who has experience selling construction-related items in large-amount orders to the person who signs the check. And believe it or not, there are plenty of candidates running around in the construction industry and closely allied fields. Such as whom, you ask? For openers, there are the factory representatives for the building material manufacturers. Factory representatives for the preengineered metal building manufacturers also fall into this category. Construction equipment salespeople also are possibilities. Any segment of the industry that uses salespeople can offer the good raw material for a construction sales job.

In the allied fields there are factory representatives for construction steel and other metal products. Industrial mill supply salespeople have an acceptable background. It's my opinion that the factory representatives for building materials and metal building manufacturers are the most suitable source of good construction sales force. They will have all the assets plus the added advantage of being used to running a sales territory on their own. To them, the home office is not across town but in the next state. To do a successful job, they have to be self-starters and able to work with very little supervision. Both of these qualities are absolutely essential to the results-getting construction salesperson. The only weakness with these representatives is their being established in your area. If you happen to find a person who's a hometown resident, then so much the better.

When you find one of these representatives who does want to make a change, do your homework to find out why. You don't want to end up with some sales manager's problem employee. You'll discover, after talking with some of these people about changing jobs, that the main reason they are thinking about leaving is the desire to get off the road and have better opportunities to make more money. This will come home to them as they grow a little older, so their age is in your favor also.

The other areas I've mentioned are good sales force sources also, only not quite as good as the factory representative group; so don't ignore these other sources. As a matter of fact, one of the best construction sales agents I ever had the pleasure of knowing and working with came from a supplier of metal products. He was a prime example of why you have to hire a salesperson first, with all else secondary. This guy was the best psychologist I've ever seen when it came to the prospect. He knew exactly what the prospect was going to say before it was said. He knew how to approach the job; better yet, he was a closer—he got the prospect's name on the dotted line. But he always had a little trouble with construction details. He never learned the correct names for all the details in some areas, such as electrical. For example, everything was a switch or a plug on the wall—never exactly what it was, simply a switch or a plug. He never could remember the difference between a column and a beam; and I'll bet right now if you asked him, he'd stop and think about it and then give you the wrong answer. After a while the two owners of the construction company that we both worked for gave up on him. If the prospects didn't mind, why should the owners? He just kept bringing in the contracts.

He was a fine salesman, always selling himself and his company and on the lookout for leads 24 hours per day, 7 days per week. This man was a super construction salesman who came from a metal products firm. He never mastered all the details on the construction side; and what's more, he would tell you he didn't care. He was a salesman, not an engineer.

My point is that no matter where you find your salesperson, she or he had better be a salesperson first.

Sales Territories Where Two or More Salespeople Are Involved

If you have or plan to have two or more sales agents, there should be definite areas of responsibility outlined. These established territories should be for leads and prospects, not actual construction sites. The territory is where the salesperson will work, seeking leads and pursuing prospects. This includes incoming phone leads from the territory as well. If the prospect is going to build and the site is in another person's territory, this situation will present no problem. The job belongs to the person who sold it. This particular situation will arise because the prospect will be moving or in some cases adding a branch operation.

There will be conflicts from time to time involving leads that start in one area and end up in another territory. My personal feeling is that the lead in such a case should be passed on to the person who handles the territory. Or you can remain uninvolved and let the two salespeople work things out. You can't afford squabbling over leads. It's very important that salespeople work together, for if you're doing any business at all, there'll be times when one has to cover for the other. Good relations in the construction sales department are a must.

You may feel the sales representatives are mature and objective enough to work out the territories on their own. If so, let them, as long as the results are there. After all, results are what you're really interested in. My last bit of advice on this subject is not to create a very formal sales department with rigid guidelines. You want an efficient, fluid operation that does what has to be done to get the contract. When I was one of two salesmen working for a construction company, a lead came to my attention that was located in the other salesman's territory. I got the lead because the party in question was a personal friend. I hesitated to pass the lead on to the other salesman because I was the one who knew the party and the background, and this person had come to me because we were friends. Therefore, I was the one with the better chance of having a contract signed. Also, it was a lead that was not out on the streets, and chances were the other salesman would never have heard of the job. We talked it over. The first thing was to get the project for the firm; and since I had the inside track, he told me to sell and handle the job, which I did. Don't shackle your salesperson with rigid rules. Let her or him operate in an independent manner, as long as it gets the results.

Salesperson Remuneration

Remuneration in plain language is money! Nothing's free, and 99 percent of the time you get exactly what you pay for. A good results-getting construction sales agent is not cheap. While hard numbers would be very difficult to present I can nevertheless offer some general rules.

I advise against a straight salary. First, if the prospective sales agent wants a straight salary, mark him or her off your list. This is a good indicator of what kind of salesperson she or he is. A straight-salaried person is not aggressive enough for construction sales. It's too easy to become lazy when the money is always there. Of course, if the agent is not working, the lack of results will show it, and you can take proper action.

Most of the salesperson's money should come from commissions, and the salesperson who is of the caliber you want will prefer it this way. Rely on your own good common business sense about income. Just remember to maintain some control, don't put a limit on earned income, and never pay a straight salary.

Now let's look at a special situation that will always come up when you have a salesperson working for you and how to handle it. Your sales agent makes money on the jobs sold so that is where he or she will be spending time. Leads and prospects are where she or he will concentrate efforts because that's where money is made. No one questions the jobs the agent seeks out and sells, but let's get into a gray area.

You, the owner of the construction company, get a phone call from a friend telling you to contact so-and-so because she is ready to add on to her metal building warehouse. You follow up and get he go-ahead to order the building and get things going. Well, to get the job going will require some time and effort, so you tell your sales agent to do this and that to get the job moving. A few weeks pass, and you are surprised to find out little has been done by the sales agent. So you ask why he or she is behind on this special job, and the agent tells you she or he has been very busy trying to sell a certain job. Plus the sales agent had a problem on one of the jobs under construction that had to be cleared up with the owner.

In other words, time was invested on jobs where he or she made money, and the sales representative was not going to put much time into running errands on somebody else's job. This can become a serious problem with the selling side of your business.

My recommendation is to include the sales force in all the company selling areas. After you got that call from your friend, you then called the prospect and said you would be sending over someone to work out the details and get things moving. Then you meet with your sales agent, give all the details, and tell him or her to sell the job. Then you can move on to running your construction company because your salesperson is in the picture and will push things along.

This is all about leads; no matter where they come from, they are passed on to your sales agent to follow up. There is no competition between the salesperson and the office. When an employee gives you a lead, then you give it to the sales force. In other words, your company is a team, and everyone works together to make sure your company gets the job.

Your salesperson works for the same reason you do, to make money; so work together, and both of you will make money from his or her efforts.

NOTES

NOTES

Appendix A
Metal Building Manufacturers Association Members

ACI Building Systems, Inc.	P.O. Box 1316 Highway 6 West Batesville, MS 38606 (601) 563-4574 Fax (601) 563-1142
Alliance Steel, Inc.	3333 South Council Road Oklahoma City, OK 73179-4410 (405) 745-7500 Fax (405) 745-7502
American Buildings Company	P.O. Box 800 Eufaula, AL 36072-9950 (334) 687-2000 Fax (334) 687-8315
American Steel Building Company, Inc.	P.O. Box 14244 Houston, TX 77221 (713) 433-5661 Fax (713) 433-0847
Behlen Mfg. Co.	P.O. Box 569 Columbus, NE 68602-0569 (402) 563-7311 Fax (402) 563-7405
Bigbee Steel Buildings, Inc.	P.O. Box 2314 Muscle Shoals, AL 35662 (205) 383-7322 Fax (205) 381-9669

Butler Manufacturing Company	P.O. Box 917 Kansas City, MO 64141-0917 (816) 968-3000 Fax (816) 968-3075
CBC, Inc.	P.O. Box 1009 1700 East Louise Avenue Lathrop, CA 95330 (209) 983-0910 Fax (209) 858-2354
Ceco Building Systems	P.O. Drawer 6500 Columbus, MS 39703-6500 (601) 328-6722 Fax (601) 327-7309
Chief Industries, Inc.	3942 Old West Highway 30 P.O. Box 2078 Grand Island, NE 68802-2078 (308) 382-8820 Fax (308) 389-7370
Crown Metal Buildings, Inc.	P.O. Box 1245 203 South 10th Street Cabot, AR 72023 (501) 843-5856
Dean Steel Buildings, Inc.	2929 Industrial Avenue Fort Myers, FL 33901 (941) 334-1051 Fax (941) 334-0932
Garco Building Systems	P.O. Box 19248 Spokane, WA 99219-9248 (509) 244-5611 Fax (509) 244-2850
Gulf States Manufacturers, L.P.	P.O. Box 1128 Starkville, MS 39760-1128 (601) 323-8021 Fax (601) 320-9642
Inland Southern Corporation	2141 Second Avenue, S.W. Cullman, AL 35055 (800) 438-1606 Fax (800) 438-1626
Kirby Building Systems, Inc.	P.O. Box 390 Kirby Drive Portland, TN 37148-0390 (615) 889-0020 Fax (615) 325-6953

Ludwig Buildings, Inc.	P.O. Box 23134 521 Timesaver Avenue Harahan, LA 70183 (504) 733-6260 Fax (504) 733-7458
Metal Building Products, Inc.	3861 Old Getwell Road Memphis, TN 38118 (901) 363-6777 Fax (901) 363-6795
Mueller, Inc.	1913 Hutchins Ballinger, TX 76821 (800) 527-1087 Fax (915) 365-8181
NCI Building Systems, Inc.	P.O. Box 40220 Houston, TX 77240-0220 (713) 466-7788 Fax (713) 849-5024
Nucor Building Systems	305 Industrial Parkway Waterloo, IN 46793 (219) 837-7891 Fax (219) 837-7384
OSI Building Systems	P.O. Box 5230 Montgomery, AL 36103 (334) 834-3500 Fax (334) 265-3850
Package Industries, Inc.	15 Harback Road Sutton, MA 01590 (508) 865-5871 Fax (508) 865-9130
Pinnacle Structures, Inc.	P.O. Box 1268 2665 Highway 321 Bypass Cabot, AR 72023 (800) 201-1534 Fax (501) 941-2675
Rigid Building Systems, Inc.	18933 Aldine Westfield Houston, TX 77073 (281) 443-9065 Fax (281) 443-9064
Ruffin Building Systems, Inc.	6914 Highway 2 Oak Grove, LA 71263-8390 (800) 421-4232 Fax (318) 428-8360

Southern Structures, Inc.	P.O. Box 52005 Lafayette, LA 70505 (318) 856-5981 Fax (318) 856-5980
Star Building Systems	P.O. Box 94910 Oklahoma City, OK 73143 (405) 636-2010 Fax (405) 636-2419
Steelox Systems, Inc.	P.O. Box 8181 Mason, OH 45040-8181 (513) 573-5200 Fax (513) 573-5299
United Structures of America, Inc.	P.O. Box 60069 Houston, TX 77205-0069 (281) 442-8247 Fax (281) 442-2115
V P Buildings	6000 Poplar Avenue Suite 400 Memphis, TN 38119 (901) 762-6000 Fax (901) 762-6008
Whirlwind Building Systems	P.O. Box 75280 Houston, TX 77234 (800) 324-9992 Fax (713) 946-5446

Appendix **B**

The BOCA® National Building Code/1996

The BOCA® *National Building Code/* 1996

Model building regulations for the protection of public health, safety and welfare.

THIRTEENTH EDITION

As recommended and maintained by the voting membership of

BUILDING OFFICIALS & CODE ADMINISTRATORS INTERNATIONAL, INC.

Serving Government & Industry Since 1915

4051 W. Flossmoor Rd.
Country Club Hills, IL 60478-5795
Telephone: 708/799-2300
Facsimile: 708/799-4981
e-mail: codes@bocai.org

REGIONAL OFFICES

Mideast Regional Office
1245 S. Sunbury Rd., Ste. 100
Westerville, Ohio 43081-9308
Telephone: 614/890-1064
Facsimile: 614/890-9712

Eastern Regional Office
One Neshaminy Interplex
Suite 201
Trevose, PA 19053-6338
Telephone: 215/638-0554
Facsimile: 215/638-4438

Southwest Regional Office
Towne Centre Complex
10830 E. 45th Street, Ste. 200
Tulsa, OK 74146-3809
Telephone: 918/664-4434
Facsimile: 918/664-4435

Northeast Regional Office
6 Omega Terrace
Latham, NY 12110-1939
Telephone: 518/782-1708
Facsimile: 518/783-0889

STRUCTURAL TESTS AND INSPECTIONS

SECTION 1701.0 GENERAL

1701.1 Scope: The provisions of this chapter shall govern the quality, workmanship and requirements for all materials hereafter used in the construction of buildings and structures. All materials of construction and tests shall conform to the applicable standards listed in this code.

1701.2 New materials: All new building materials, equipment, appliances, systems or methods of construction not provided for in this code, and any material of questioned suitability proposed for use in the construction of a building or structure, shall be subjected to the tests prescribed in this chapter and in the *approved rules* to determine character, quality and limitations of use.

1701.3 Used materials: The use of all second-hand materials which meet the minimum requirements of this code for new materials shall be permitted.

SECTION 1702.0 DEFINITIONS

1702.1 General: The following words and terms shall, for the purposes of this chapter and as used elsewhere in this code, have the meanings shown herein.

Approved agency: An established and recognized agency regularly engaged in conducting tests or furnishing inspection services, when such agency has been approved (see Section 1704.0).

Fabricated item: Structural, loadbearing or lateral *load*-resisting assemblies consisting of materials assembled prior to installation in a building or structure, or subjected to operations such as heat treatment, thermal cutting, cold working or reforming after manufacture and prior to installation in a building or structure. Materials produced in accordance with standard specifications referenced by this code, such as rolled structural steel shapes, steel-reinforcing bars, masonry units and plywood sheets, shall not be considered "fabricated items."

Inspection certificate: An identification applied on a product by an approved agency containing the name of the manufacturer, the function and performance characteristics, and the name and identification of an approved agency which indicates that the product or material has been inspected and evaluated by an approved agency (see Section 1704.3 and also Mark, Manufacturer's designation and Label).

Inspection, special: Inspection as herein required of the installation, fabrication, erection or placement of components and connections requiring special expertise to ensure adequacy (see Sections 114.0 and 1705.0).

Label: An identification applied on a product by the manufacturer which contains the name of the manufacturer, the function and performance characteristics of the product or material, and the name and identification of an approved agency and which indicates that the representative sample of the product or material has been tested and evaluated by an approved agency (see Section 1704.3 and also Mark, Manufacturer's designation and Inspection certificate).

Manufacturer's designation: An identification applied on a product by the manufacturer indicating that a product or material complies with a specified standard or set of rules (see also Mark, Label and Inspection certificate.)

Mark: An identification applied on a product by the manufacturer indicating the name of the manufacturer and the function of a product or material (see also Manufacturer's designation, Label and Inspection certificate.)

SECTION 1703.0 INFORMATION REQUIRED

1703.1 Material performance: Where the quality of materials is essential for conformance to this code, specific information shall be given to establish such quality; and this code shall not be cited, or the term "legal" or the term's equivalent be used as a substitute for specific information. This information shall consist of test reports conducted by an *approved testing agency* in accordance with the standards referenced in Chapter 35, or other such information as necessary for the code official to determine that the material meets the applicable code requirements.

1703.1.1 Labeling: Where materials or assemblies are required by this code to be *labeled*, such materials and assemblies shall be *labeled* by an *approved agency* in accordance with Section 1704.0.

1703.2 Research and investigation: Sufficient technical data shall be submitted to substantiate the proposed use of any material or assembly. If it is determined that the evidence submitted is satisfactory proof of performance for the use intended, the code official shall approve the use of the material or assembly subject to the requirements of this code. The cost of all tests, reports and investigations required under these provisions shall be paid by the permit applicant.

1703.2.1 Research reports: Supporting data, where necessary to assist in the approval of all materials or assemblies not specifically provided for in this code, shall consist of valid research reports from approved sources.

1703.3 Evaluation and follow-up inspection services: Prior to the approval of a closed prefabricated assembly, the permit applicant shall submit an evaluation report of each prefabricated assembly. The report shall indicate the complete details of the assembly, including a description of the assembly and the assembly's components, the basis upon which the assembly is being evaluated, test results and similar information, and other data as necessary for the code official to determine conformance to this code.

1703.3.1 Evaluation service: The code official shall review evaluation reports from approved sources for adequacy and conformance to the code.

1703.3.2 Follow-up inspection: The owner shall provide for *special inspections* of *fabricated items* in accordance with Section 1705.2.

1703.3.3 Test and inspection records: Copies of all necessary test and inspection records shall be filed with the code official.

SECTION 1704.0 APPROVALS

1704.1 Written approval: Any material, appliance, equipment, system or method of construction meeting the requirements of this code shall be approved in *writing* within a reasonable time after satisfactory completion of all the required tests and submission of required test reports.

1704.2 Approved record: For any material, appliance, equipment, system or method of construction that has been approved, a record of such approval, including all of the conditions and limitations of the approval, shall be kept on file in the code official's office and shall be open to public inspection at all appropriate times.

1704.3 Labeling: Products and materials required to be *labeled* shall be *labeled* in accordance with the procedures set forth in Sections 1704.3.1 through 1704.3.3.

1704.3.1 Testing: An *approved agency* shall test a representative sample of the product or material being *labeled* to the relevant standard or standards. The *approved agency* shall maintain a record of all of the tests performed. The record shall provide sufficient detail to verify compliance with the test standard.

1704.3.2 Inspection and identification: The *approved agency* shall periodically perform an inspection, which shall be in-plant if necessary, of the product or material that is to be *labeled*. The inspection shall verify that the *labeled* product or material is representative of the product or material tested.

1704.3.2.1 Independent: The *agency* to be approved shall be objective and competent. The *agency* shall also disclose all possible conflicts of interest so that objectivity can be confirmed.

1704.3.2.2 Equipment: An *approved agency* shall have adequate equipment to perform all required tests. The equipment shall be periodically calibrated.

1704.3.2.3 Personnel: An *approved agency* shall employ experienced personnel educated in conducting, supervising and evaluating tests.

1704.3.3 Label information: The *label* shall contain the manufacturer's or distributor's identification, model number, serial number, or definitive information describing the product or material's performance characteristics and *approved agency's* identification.

1704.4 Heretofore-approved materials: The use of any material already *fabricated* or of any construction already erected, which conformed to requirements or approvals heretofore in effect, shall be permitted to continue, if not detrimental to life, health or safety of the public.

SECTION 1705.0 SPECIAL INSPECTIONS

1705.1 General: The permit applicant shall provide *special inspections* where application is made for construction as described in this section. The special inspectors shall be provided by the permit applicant and shall be qualified and approved for the inspection of the work described herein.

Exceptions

1. *Special inspections* are not required for work of a minor nature or where warranted by conditions in the jurisdiction.
2. *Special inspections* are not required for building components unless the design involves the practice of professional engineering or architecture as defined by applicable state statutes and regulations governing the professional registration and certification of engineers or architects.
3. *Special inspections* are not required for occupancies in Use Group R-3 and occupancies in Use Group U that are accessory to a residential occupancy including, but not limited to, those listed in Table 312.1.

1705.1.1 Building permit requirement: The permit applicant shall submit a statement of *special inspections* prepared by the registered design professional in responsible charge in accordance with Section 114.2.1 as a condition for permit issuance. This statement shall include a complete list of materials and work requiring *special inspection* by this section, the *inspections* to be performed and a list of the individuals, *approved agencies* and firms intended to be retained for conducting such inspections.

1705.1.2 Report requirement: Special inspectors shall keep records of all *inspections*. The special inspector shall furnish *inspection* reports to the code official, and to the *registered design professional* in responsible charge. All discrepancies shall be brought to the immediate attention of the contractor for correction. If the discrepancies are not corrected, the discrepancies shall be brought to the attention of the code official and to the *registered design professional* in responsible charge prior to the completion of that phase of the work. A final report of *inspections* documenting completion of all required *special inspections* and correction of any discrepancies noted in the *inspections* shall be submitted prior to the issuance of a certificate of occupancy. Interim reports shall be submitted periodically at a frequency agreed upon by the permit applicant and the code official prior to the start of work.

1705.2 Inspection of fabricators: Where fabrication of structural loadbearing members and assemblies is being performed on the premises of a fabricator's shop, *special inspection* of the *fabricated items* shall be required. The *fabricated items* shall be *inspected* as required by this section and as required elsewhere in this code.

1705.2.1 Fabrication procedures: The special inspector shall verify that the fabricator maintains detailed fabrication and quality control procedures which provide a basis for inspection control of the workmanship and the fabricator's ability to conform to approved drawings, project specifications and referenced standards. The special inspector shall review the procedures for completeness and adequacy relative to the code requirements for the fabricator's scope of work.

1705.2.2 Procedures implementation: The special inspector shall verify that the fabricator is properly implementing the fabrication and quality control procedures outlined in Section 1705.2.1.

Exception: *Special inspections* as required by Section 1705.2 shall not be required where the fabricator maintains an agreement with an *approved independent inspection or quality control agency* to conduct periodic in-plant *inspections* at the fabricator's plant, at a frequency that will assure the fabricator's conformance to the requirements of the *inspection agency's* approved quality control program.

1705.3 Steel construction: The *special inspections* for steel elements of buildings and structures shall be as required by Sections 1705.3.1 through 1705.3.3.

1705.3.1 Inspection of steel fabricators: The permit applicant shall provide *special inspection* of steel *fabricated items* in accordance with the provisions of Section 1705.2.

Exception: *Special inspection* of the steel fabrication process shall not be required where the fabricator does not perform any welding, thermal cutting or heating operation of any kind as part of the fabrication process. In such cases, the fabricator shall be required to submit a detailed procedure for material control which demonstrates the fabricator's ability to maintain suitable records and procedures such that, at any time during the fabrication process, the material specification, grade and mill test reports for the main stress-carrying elements and bolts are capable of being determined.

1705.3.2 Material receiving: All main stress-carrying elements, welding material and bolting material shall be *inspected* for conformance to Table 1705.3.2.

1705.3.3 Erection: *Special inspections* are required for bolts, welding and details as specified in Sections 1705.3.3.1 through 1705.3.3.3.

1705.3.3.1 Installation of high-strength bolts: *Inspection* shall be as specified in Section 9 of the RCSC *Specification for Structural Joints Using A325 or A490 Bolts* listed in Chapter 35.

Table 1705.3.2
INSPECTION FOR STEEL MATERIALS

Material	Inspection required	Reference[a] for criteria
Bolts, nuts, washers	1. Material identification markings. 2. Conformance to ASTM standards specified by the design engineer. Manufacturer's designation (certificate of compliance) is required.	Applicable ASTM material specifications; AISC ASD, Section A3.4; AISC LRFD, Section A3.3
Structural steel	1. Material identification markings. 2. Conformance to ASTM standards specified in the approved plans and specifications.	ASTM A6 or ASTM A588 Provide certified test reports in accordance with ASTM A6 or ASTM A588
Weld filler materials	1. Conformance to AWS specification as specified in the approved plans and specifications. Manufacturer's designation (certificate of compliance) is required.	AISC ASD, Section A3.6; AISC LRFD, Section A3.5

Note a. The specific standards referenced are those listed in Chapter 35.

1705.3.3.2 Welding: Weld *inspection* shall be in compliance with Section 6 of AWS D1.1 listed in Chapter 35. Weld inspectors shall be certified in accordance with AWS D1.1 listed in Chapter 35.

1705.3.3.2.1 Welding of the structural seismic-resisting system: Welding of the structural seismic-resisting system of buildings assigned to Seismic Performance Category C, D or E, in accordance with Section 1610.1.7, shall be inspected in accordance with Sections 1705.3.3.2.2 and 1705.3.3.2.3. Each complete penetration groove weld in joints and splices shall be tested for the full length of the weld either by ultrasonic testing or by other approved methods, for special moment frames and eccentrically braced frames.

Exception: The nondestructive testing rate for welds made by an individual welder is permitted to be reduced to 25 percent of the welds, with the approval of the *registered design professional* responsible for the structural design, provided the weld inspection reject rate is 5 percent or less.

1705.3.3.2.2 Column splice welds: Column splice welds, which are partial penetration groove welds, shall be tested by ultrasonic testing or other approved methods at a percentage rate established by the *registered design professional* responsible for the structural design. All partial penetration column splice welds designed for axial or flexural tension from seismic forces shall be tested.

1705.3.3.2.3 Base metal testing: Base metal having a thickness more than $1^1/_2$ inches (38 mm) and subject to through-thickness weld shrinkage strains shall be ultrasonically tested for discontinuities behind and adjacent to the welds after joint welding. Any material discontinuities shall be evaluated based on the criteria established in the *construction documents* by the *registered design professional* responsible for the structural design.

1705.3.3.3 Details: The special inspector shall perform an *inspection* of the steel frame to verify compliance with the details shown on the approved *construction documents*, such as bracing, stiffening, member locations and proper application of joint details at each connection.

1705.4 Concrete construction: The *special inspections* for concrete elements of buildings and structures and concreting operations shall be as required by Sections 1705.4.1 through 1705.4.7.

Exception: *Special inspections* shall not be required for:

1. Concrete footings of buildings three stories or less in height which are fully supported on earth or rock.
2. Nonstructural concrete slabs supported directly on the ground, including prestressed slabs on grade, where the effective prestress in the concrete is less than 150 psi (0.11 kg/mm^2).
3. Plain concrete foundation walls constructed in accordance with Table 1812.3.2.
4. Concrete patios, driveways and sidewalks, on grade.

1705.4.1 Materials: In the absence of sufficient data or documentation providing evidence of conformance to quality standards for materials in Chapter 3 of ACI 318 listed in Chapter 35, the code official shall require testing of materials in accordance with the appropriate standards and criteria for the material in Chapter 3 of ACI 318 listed in Chapter 35. Weldability of reinforcement, except that which conforms to ASTM A706 listed in Chapter 35, shall be determined in accordance with the requirements of Section 1906.5.2.

1705.4.2 Installation of reinforcing and prestressing steel: The location and installation details of reinforcing and prestressing steel shall be *inspected* for compliance with the approved *construction documents* and ACI 318 (such as Sections 7.4, 7.5, 7.6 and 7.7) listed in Chapter 35. Welding of reinforcing of the structural seismic-resisting system shall be inspected for buildings assigned to Seismic Performance Category C, D or E, in accordance with Section 1610.1.7.

1705.4.3 Formwork: Forms for concrete, if used, shall be *inspected* for compliance with Section 6.1 of ACI 318 listed in Chapter 35, and with any additional design requirements indicated on the approved *construction documents*. *Inspection* of form removal and reshoring shall be conducted to verify compliance with Section 6.2 of ACI 318 listed in Chapter 35.

1705.4.4 Concreting operations: During placing and curing of concrete, the *special inspections* listed in Table 1705.4.4 shall be performed.

1705.4.5 Inspection during prestressing: *Inspection* during the application of prestressing forces shall be performed to determine compliance with Section 18.18 of ACI 318 listed in Chapter 35.

1705.4.5.1 Inspection during grouting: In buildings assigned to Seismic Performance Category C, D or E, in accordance with Section 1610.1.7, inspection during the grouting of bonded prestressing tendons in the structural seismic-resisting system shall be performed.

Table 1705.4.4
REQUIRED INSPECTIONS DURING CONCRETING

Required inspection	Reference[a] for criteria
1. Evaluation of concrete strength, except as exempted by Section 1908.3.1(3) of this code.	ACI 318 Section 5.6
2. Inspection for use of proper mix proportions and proper mix techniques.	ACI 318 Chapter 4, Sections 5.2, 5.3, 5.4 and 5.8
3. Inspection during concrete placement, for proper application techniques.	ACI 318 Sections 5.9 and 5.10
4. Inspection for maintenance of specified curing temperatures and techniques.	ACI 318 Sections 5.11, 5.12 and 5.13

Note a. ACI 318 listed in Chapter 35.

1705.4.6 Manufacture of precast concrete: The manufacture of precast concrete, as required by Section 1705.2, shall be subject to a quality control program administered by an *approved agency*.

1705.4.7 Erection of precast concrete: Erection of precast concrete shall be *inspected* for compliance with the approved plans and erection drawings.

1705.5 Masonry construction: The *special inspections* listed in Table 1705.5 shall be required for masonry construction where masonry is designed in accordance with ACI 530/ASCE 5/TMS 402 listed in Chapter 35.

Table 1705.5
SPECIAL INSPECTIONS FOR MASONRY CONSTRUCTION

Inspection or test	Referenced[a] criteria	
	ACI 530/ ASCE 5/ TMS 402	ACI 530.1/ ASCE 6/ TMS 602
1. Material		Sec. 2.3
2. Masonry strength		Sec. 1.4
3. Construction operations:		
a. Proportioning, mixing consistency of mortar and grout		Sec. 2.6
b. Application of mortar and grout; installation of masonry units		Sec. 3.2 Sec. 3.5
c. Condition, size, location and spacing of reinforcement	Chapter 8	
d. Protection of masonry during cold weather (temperature below 40 degrees F.) or hot weather (temperature above 100 degrees F.)		
e. Anchorage	Sec. 4.2 Sec. 5.14	Sec. 1.8
4. Inspection of welding of reinforcement, grouting, consolidation and reconsolidation for buildings assigned to Seismic Performance Category C, D or E, in accordance with Section 1610.1.7.	Note b	Note b

Note a. The specific standards referenced are those listed in Chapter 35.
Note b. Referenced criteria not applicable.

1705.6 Wood construction: *Special inspections* of the fabrication process of wood structural elements and assemblies shall be in accordance with Section 1705.2. *Special inspection* is required for nailing, bolting, structural gluing or other fastening of the structural seismic-resisting system of buildings assigned to Seismic Performance Category C, D or E, in accordance with Section 1610.1.7.

1705.7 Prepared fill: The *special inspections* for prepared fill shall be as required by Sections 1705.7.1 through 1705.7.3. The approved report, required by Section 1804.1, shall be used to determine compliance.

1705.7.1 Site preparation: Prior to placement of the prepared fill, the special inspector shall determine that the site has been prepared in accordance with the approved report.

1705.7.2 During fill placement: During the placement and compaction of the fill material, the special inspector shall determine that the material being used and the maximum lift thicknesses comply with the approved report.

1705.7.3 Evaluation of in-place density: The special inspector shall determine, at the approved frequency, that the in-place dry density of the compacted fill complies with the approved report.

1705.8 Pile foundations: *Special inspections* of pile foundations are required as provided for in Section 1816.13 of this code.

1705.9 Pier foundations: *Special inspection* is required for pier foundations of buildings assigned to Seismic Performance Category C, D or E, in accordance with Section 1610.1.7.

1705.10 Wall panels and veneers: *Special inspection* is required for exterior and interior architectural wall panels and the anchoring of veneers for buildings assigned to Seismic Performance Category E, in accordance with Section 1610.1.7.

1705.11 Mechanical and electrical components: Mechanical and electrical components that are located in buildings assigned to Seismic Performance Category E shall be inspected, tested and certified as required by this section, in accordance with Section 1610.1.7.

1705.11.1 Component inspection: *Special inspection* is required for the installation of the following components where the component has a performance criteria factor of 1.0 or 1.5 in accordance with Section 1610.6.4.

1. Equipment using combustible energy sources.
2. Electrical motors, transformers, switchgear unit substations and motor control centers.
3. Reciprocating and rotating-type machinery.
4. Piping distribution systems, 3 inches and larger.
5. Tanks, heat exchangers and pressure vessels.

1705.11.2 Component and attachment testing: The component manufacturer shall test or analyze the component and the component mounting system or anchorage for the design forces in Section 1610.6.4 for those components having a performance criteria factor of 1.0 or 1.5 in accordance with Section 1610.6.4. The manufacturer shall submit a certificate of compliance for review and acceptance by the *registered design professional* responsible for the design, and for approval by the code official. The basis of certification shall be by test on a shaking table, by three-dimensional shock tests, by an analytical method using dynamic characteristics and forces from Section 1610.6.4 or by more rigorous analysis. The special inspector shall inspect the component and verify that the *label*, anchorage or mounting conform to the certificate of compliance.

1705.11.3 Component manufacturer certification: Each manufacturer of equipment to be placed in a building assigned to Seismic Performance Category E, in accordance with Section 1610.1.7, where the equipment has a performance criteria factor of 1.0 or 1.5 in accordance with Section 1610.6.4, shall maintain an approved quality control program. Evidence of the quality control program shall be permanently identified on each piece of equipment by a label.

1705.12 Sprayed cementitious and mineral fiber fireresistive materials: Special inspections for sprayed cementitious and mineral fiber fireresistive materials applied to structural elements and decks shall be in accordance with Sections 1705.12.1 through 1705.12.5. Special inspections shall be based upon the fireresistance design as designated in the approved construction documents.

1705.12.1 Structural member surface conditions: The surfaces shall be prepared in accordance with the approved fireresistance design and the approved manufacturer's written instructions. The prepared surface of all structural members to be sprayed shall be inspected before the application of the sprayed fireresistive material.

1705.12.2 Application: The substrate shall have a minimum ambient temperature before and after application as specified in the approved manufacturer's written instructions. The area for application shall be ventilated during and after application as required by the approved manufacturer's written instructions.

1705.12.3 Thickness: The thickness of the sprayed cementitious and mineral fiber fireresistive materials applied to structural elements shall not be less than the thicknesses required by the approved fireresistance design. Thickness shall be determined by an approved method. Samples of the sprayed cementitious and mineral fiber fireresistive materials shall be selected in accordance with Sections 1705.12.3.1 and 1705.12.3.2.

1705.12.3.1 Floor, roof and wall assemblies: The thickness of the sprayed cementitious and mineral fiber fireresistive material applied to the underside of floor, roof and wall assemblies shall be determined by taking the average of ten measurements in a 144-square-inch (0.093 m^2) sample area, having a minimum width of 6 inches (152 mm), for each 1,000 square feet (93 m^2) of the sprayed area on each floor or part thereof.

1705.12.3.2 Structural frame members: The thickness of the sprayed cementitious and mineral fiber fireresistive material applied to structural members shall be determined by taking nine measurements at a single cross section for structural frame beams or girders, seven measurements at a single cross section for joists and trusses and twelve measurements at single cross section for columns. Thickness testing shall be performed on 25 percent of the structural members on each floor.

1705.12.4 Density: The density of the sprayed cementitious and mineral fiber fireresistive material shall not be less than the density specified in the approved fireresistance design. Density of the sprayed cementitious and mineral fiber fireresistive material shall be determined by an approved method at the frequency specified in Sections 1705.12.3.1 and 1705.12.3.2.

1705.12.5 Bond strength: The cohesive/adhesive bond strength of the cured sprayed cementitious and mineral fiber fireresistive material applied to structural elements shall not be less than the cohesive/adhesive bond strength specified in the approved fireresistance design. The cohesive/adhesive bond strength shall be determined by taking samples of the sprayed cementitious and mineral fiber fireresistive material as specified in Sections 1705.12.3.1 and 1705.12.3.2.

1705.13 Exterior insulation and finish systems (EIFS): Special inspections are required for field application of EIFS and prefabricated EIFS panels (see Section 1405.8).

1705.14 Special cases: *Special inspections* shall be required for proposed work which is, in the opinion of the code official, unusual in its nature, such as:

1. Construction of materials and systems which are alternatives to materials and systems prescribed by this code.
2. Unusual design applications of materials described in this code.
3. Materials and systems required to be installed in accordance with additional manufacturer's instructions that prescribe requirements not contained in this code or in standards referenced by this code.

SECTION 1706.0 DESIGN STRENGTHS OF MATERIALS

1706.1 Conformance to standards: The design strengths and permissible stresses of any structural material that is identified by a manufacturer's designation as to manufacture and grade by mill tests, or the strength and stress grade is otherwise confirmed to the satisfaction of the code official, shall conform to the specifications and methods of design of accepted engineering practice or the *approved rules* in the absence of applicable standards.

1706.2 New materials: For materials which are not specifically provided for in this code, the design strengths and permissible stresses shall be established by tests as provided for in Sections 1708.0 and 1709.0.

SECTION 1707.0 ALTERNATIVE TEST PROCEDURE

1707.1 General: In the absence of *approved rules* or other approved standards, the code official shall make, or cause to be made, the necessary tests and investigations; or the code official shall accept duly authenticated reports from *approved agencies* in respect to the quality and manner of use of new materials or assemblies as provided for in Section 106.0. The cost of all tests and other investigations required under the provisions of this code shall be borne by the permit applicant.

SECTION 1708.0 TEST SAFE LOAD

1708.1 Where required: Where proposed construction is not capable of being designed by approved engineering analysis, or where proposed construction design method does not comply with the applicable material design standard listed in Chapter 35, the system of construction or the structural unit and the connections shall be subjected to the tests prescribed in Section 1710.0. The code official shall accept certified reports of such tests conducted by an *approved testing agency*, provided that such tests meet the requirements of this code and approved procedures.

SECTION 1709.0 IN-SITU LOAD TESTS

1709.1 General: Whenever there is a reasonable doubt as to the stability or loadbearing capacity of a completed building, structure or portion thereof for the expected *loads*, an engineering assessment shall be required. The engineering assessment shall involve either a structural analysis or an in-situ load test, or both. The structural analysis shall be based upon actual material properties and other as-built conditions which affect stability or loadbearing capacity, and shall be conducted in accordance with the applicable design standard listed in Chapter 35. If the structural assessment determines that the loadbearing capacity is less than that required by the code, load tests shall be conducted in accordance with Section 1709.2. If the building, structure or portion thereof is found to have inadequate stability or loadbearing capacity for the expected *loads*, modifications to ensure structural adequacy or the removal of the inadequate construction shall be required.

1709.2 Test standards: All structural components and assemblies shall be tested in accordance with the appropriate material standards listed in Chapter 35. In the absence of a standard listed in Chapter 35 that contains an applicable load test procedure, the test procedure shall be developed by a *registered design professional* and approved. The test procedure shall simulate the *loads* and conditions of application that the completed structure or portion thereof will be subjected to in normal use.

1709.3 In-situ load tests: All in-situ load tests shall be conducted in accordance with Section 1709.3.1 or 1709.3.2 and shall be supervised by a *registered design professional*. The test shall simulate the applicable *loading* conditions specified in Chapter 16 as necessary to address the concerns regarding structural stability of the building, structure or portion thereof.

1709.3.1 Load test procedure specified: Where a standard listed in Chapter 35 contains an applicable load test procedure and acceptance criteria, the test procedure and acceptance criteria in the standard shall apply. In the absence of specific *load* factors or acceptance criteria, the *load* factors and acceptance criteria in Section 1709.3.2 shall apply.

1709.3.2 Load test procedure not specified: In the absence of applicable load test procedures contained within a standard referenced by this code or acceptance criteria for a specific material or method of construction, such existing structure shall be subjected to a test load equal to two times the design *load*. The test load shall be left in place for a period of 24 hours. The structure shall be considered to have met successfully the test requirements if all of the following criteria are satisfied:

1. Under the design *load*, the deflection shall not exceed the limitations specified in Section 1604.5;

2. Within 24 hours after removal of the test load, the structure shall have recovered not less than 75 percent of the maximum deflection; and
3. During and immediately after the test, the structure shall not show evidence of failure.

SECTION 1710.0 PRECONSTRUCTION LOAD TESTS

1710.1 General: In evaluating the physical properties of materials and methods of construction which are not capable of being designed by approved engineering analysis or which do not comply with the applicable material design standards listed in Chapter 35, the structural adequacy shall be predetermined based on the load test criteria established by Sections 1710.2 through 1710.5.

1710.2 Load test procedures specified: Where specific load test procedures, *load* factors and acceptance criteria are included in the applicable design standards listed in Chapter 35, such test procedures, *load* factors and acceptance criteria shall apply. In the absence of specific test procedures, *load* factors or acceptance criteria, the corresponding provisions in Section 1710.3 shall apply.

1710.3 Load test procedures not specified: Where load test procedures are not specified in the applicable design standards listed in Chapter 35, the loadbearing capacity of structural components and assemblies shall be determined on the basis of load tests conducted in accordance with Sections 1710.3.1 and 1710.3.2. Load tests shall simulate all of the applicable *loading* conditions specified in Chapter 16.

1710.3.1 Test procedure: The test assembly shall be subjected to an increasing superimposed load equal to not less than two times the superimposed design *load*. The test load shall be left in place for a period of 24 hours. The tested assembly shall be considered to have met successfully the test requirements if the assembly recovers not less than 75 percent of the maximum deflection within 24 hours after the removal of the test load. The test assembly shall then be reloaded and subjected to an increasing superimposed load until either structural failure occurs or the superimposed load is equal to two and one-half times the load at which the deflection limitations specified in Section 1710.3.2 were reached, or the load is equal to two and one-half times the superimposed design *load*. In the case of structural components and assemblies for which deflection limitations are not specified in Section 1710.3.2, the test specimen shall be subjected to an increasing superimposed load until structural failure occurs or the load is equal to two and one-half times the desired superimposed design *load*. The allowable superimposed design *load* shall be taken as the lesser of:

1. The load at the deflection limitation given by Section 1710.3.2;
2. The failure load divided by 2.5; or
3. The maximum load applied divided by 2.5.

1710.3.2 Deflection: The deflection of structural members under the design *load* shall not exceed the limitations in Section 1604.5.

1710.4 Wall and partition assemblies: Loadbearing wall and partition assemblies shall sustain the test load both with and without window framing. The test load shall include all design *load* components.

1710.5 Test specimens: All test specimens and construction shall be representative of the materials, workmanship and details normally used in practice. The properties of the materials used to construct the test assembly shall be determined on the basis of tests on samples taken from the load test assembly or on representative samples of the materials used to construct the load test assembly. All required tests shall be conducted or witnessed by an *approved agency*. Wall and partition assemblies shall be tested both with and without door and window framing.

Appendix C
Certified Excellence

CERTIFIED EXCELLENCE

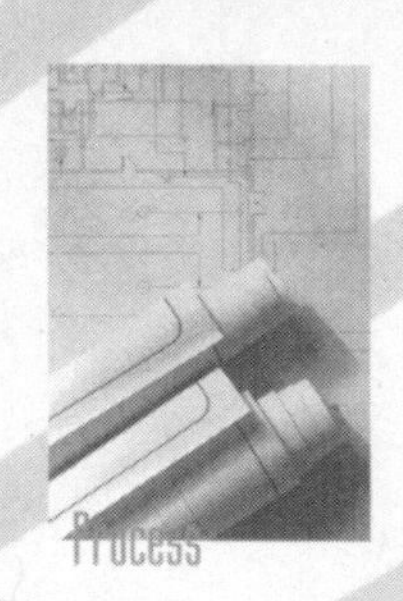

THE AMERICAN INSTITUTE OF STEEL CONSTRUCTION'S QUALITY CERTIFICATION PROGRAM FOR METAL BUILDING SYSTEMS MANUFACTURERS

Independent organizations that establish quality standards have become an important part of the professional landscape in the United States. Attorneys, for example, have the American Bar Association, and accountants have the American Institute of Certified Public Accountants. In the metal building systems industry, the American Institute of Steel Construction offers a comprehensive certification program that sets exacting professional standards.

THE PROGRAM

The AISC Quality Certification program is designed to provide a uniform, nationally recognized certification process for metal building systems manufacturers. To maintain a high level of credibility, the program includes a rigorous examination of a participant's engineering and manufacturing policies and procedures, as well as its quality assurance and control standards.

The program is rapidly gaining popularity with both private industry and government agencies. Savvy design and construction professionals are incorporating AISC certification directly into their design contracts and specifications by requiring that "The manufacturing company shall be certified under the American Institute of Steel Construction's Quality Certification program for metal building systems manufacturers." This requirement ensures that the manufacturer has achieved a high level of competency in all aspects of design and fabrication.

THE PROCESS

The certification program examines policies and procedures at each of the manufacturer's facilities and, on a limited basis, the application of those policies and procedures to randomly selected projects. It consists of nearly 200 questions covering the areas of general management, engineering and drafting, procurement, manufacturing and quality control.

Inspection and evaluation teams from a prestigious independent engineering auditing firm annually observe and evaluate the manufacturer in almost every aspect of operation. In addition, unannounced and unscheduled inspections verify continuous compliance with the detailed certification standards

CRITICAL CRITERIA FOR CERTIFICATION

Eight of the most critical criteria for achieving AISC Quality Certification include:

1. **The management team must fully communicate its commitment to achieving ongoing quality certification in accordance with all program requirements.**
2. **All submittal documentation must be in place — including job descriptions, policies and procedures.**
3. **Finished products in the plant must conform with design and quality parameters.**
4. **All engineering software must incorporate only those material specifications currently approved for metal building systems.**
5. **Raw materials in the plant must match engineering specifications.**
6. **Welding personnel and/or procedures must be properly documented, qualified and/or certified, and plant welding practices must conform with American Welding Society (AWS) procedures and workmanship requirements.**
7. **All building code requirements must be met or exceeded.**
8. **A quality control/quality assurance procedure must be articulated and practiced.**

established by AISC, which has been setting steel construction standards and writing specifications for more than 75 years. These auditors verify that the manufacturer has a successful, up to date and active quality assurance program in place.

As a result of all these measures, architects, design professionals, building code officials and the insurance industry can be assured that certified metal building systems manufacturers are capable of meeting the industry's highest standards for product and design integrity.

THE ELEMENTS

AISC Quality Certification monitors and encourages the use of sound structural engineering principles in engineering software development, design and detailing practices, and order screening procedures. This promotes accurate translation of recognized building codes into detail designs, drawings, instructions and procedures. Manufacturers use recognized engineering design methods, incorporate changes in accepted standards, and provide extensive and active continuing education programs to maintain their certification.

The certification program calls for a random sampling of the manufacturer's procedures for such items as welding, painting and bolt installation, as well as compliance with applicable building codes. It also includes random spot reviews of raw materials (including certified mill reports) and the product — from the work-in-process stage through manufacturing of the final product. All materials and services must conform to engineering documents, specifications and design requirements.

QUALITY UNDER THE GLASS

The certification process sets exacting standards for and closely scrutinizes the manufacturer's management, engineering and manufacturing processes.

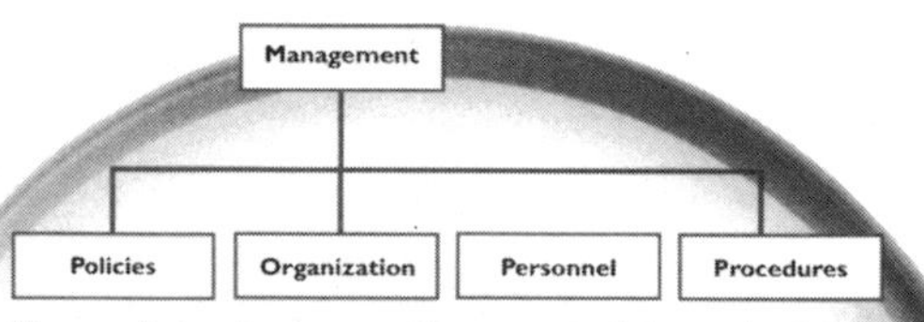

The manufacturer's management team must be fully committed to the certification process in order to achieve optimum results.

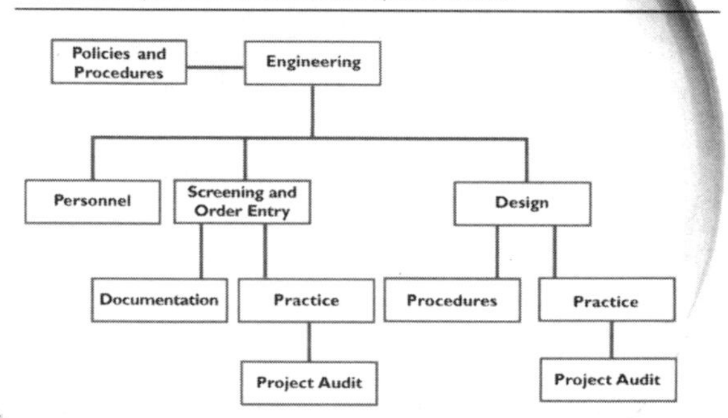

Certification monitors and encourages the use of sound structural principles in all aspects of the engineering process.

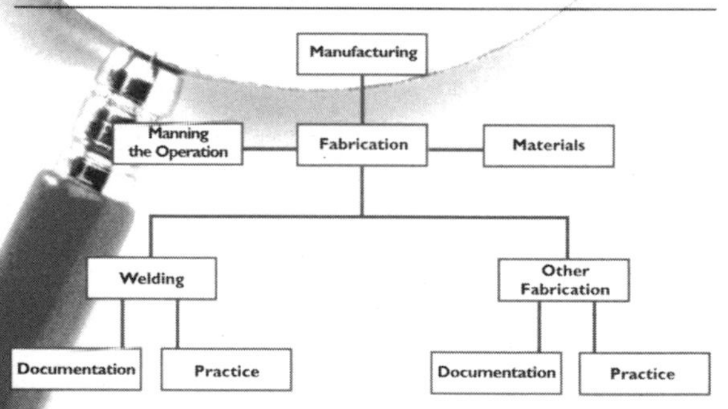

Metal building systems manufacturers are evaluated on specific manufacturing processes, as well as the materials being produced.

THE BENEFITS

These are some of the benefits that building owners, architects, specifiers and building code officials derive from the AISC Quality Certification program:

- Certified manufacturers have undergone rigorous third-party examination of their engineering and manufacturing policies, procedures and practices.
- Quality assurance standards and controls meet the requirements established in the certification program.
- Unannounced annual on-site audits show evidence of continued compliance with the program requirements.
- Certified manufacturers have demonstrated design and quality assurance procedures and practices capable of producing metal building systems that meet the needs of predictable structural integrity and quality.
- Local, national and international code groups can utilize an already established and nationally recognized certification agency to verify compliance with their standards.

Certified companies are entitled to display the AISC Quality Certification program logo for metal building systems manufacturers, underscoring their commitment to quality in all areas of design, construction, materials and management integrity. This logo also confirms that the manufacturer has met the industry's most rigorous certification standards.

Manufacturers earning AISC Quality Certification receive a certificate valid for three years. This certificate is renewed annually when the manufacturer meets all program criteria.

Becoming certified is a difficult task, as anyone familiar with the process can attest. But for everyone involved in metal building systems construction, the end result provides numerous tangible benefits.

To obtain additional information about the AISC Quality Certification program for metal building systems manufacturers or a list of current certified members, contact AISC at (312) 670-5435 or visit its Internet home page at http://www.aiscweb.com.

Metal Building Manufacturers Association
1300 Sumner Avenue
Cleveland, OH 44115-2851
Phone (216) 241-7333
FAX (216) 241-0105
Internet Address: http://www.taol.com/mbma
E-Mail Address: mbma@taol.com

American Institute of Steel Construction, Inc.
One East Wacker Drive, Suite 3100
Chicago, IL 60601-2001
Phone (312) 670-5435
FAX (312) 670-5403
Internet Address: http://www.aiscweb.com
E-Mail Address: aiscengr@dial.cic.net

Appendix D

Modern Steel Construction

MODERN STEEL
CONSTRUCTION
AISC
November 1997
$3.00

Metal Building Systems: The Low-Rise Solution

Today's metal buildings are much more attractive and versatile than their more industrial forebears

By W. Lee Shoemaker, Ph.D., P.E.

The metal building systems industry has been an innovative leader in low-rise construction. Because the industry has grown and matured so rapidly, many misconceptions exist that are based on earlier predecessors of today's advanced systems. An entire article could easily be devoted to the historical evolution of this industry that has pioneered many practices that are becoming more widely utilized. The metal building industry was first in recognizing the advantage of combining engineering and fabrication to more effectively provide an optimally designed steel framed building. Also, the movement toward design/build has been an integral part of the metal building system industry since its inception.

The focus of this article is the engineered product furnished by metal building manufacturers that is the basis for the growth in their market share. Of the low-rise, non-residential buildings under 150,000 sq. ft., metal building systems are selected by building owners and architects in nearly 70% of new construction (as referenced in the *1996 Business Review*, Metal Building Manufacturers Association). This high market share is due, in no small part, to the engineering and quality assurance that contribute to the excellent performance of these buildings.

The Metal Building Manufacturers Association (MBMA) celebrated its 40th anniversary last year. Composed of 31 Manufacturer Members and 36 Associate Members, MBMA has taken the industry lead over the years in setting the standard for excellence. In fact, MBMA has decided to make AISC Certification a requirement for membership, effective in the year 2000. MBMA's Manufacturer Members are also Associate Members of AISC.

What is a Metal Building System?

Many outside the industry still harbor the notion that a metal building is selected from a catalog of standard designs, based on the size of the building.

In fact, most metal building manufacturers custom design a building by order to within 1/16" in any dimension, based on the building code in effect, the loading conditions, and material specified. MBMA member companies have registered professional engineers on staff who are highly skilled and apply sound engineering principles to the optimal design of metal building systems. Advanced computer methods are used to help facilitate this design customization and optimization. This is also why the industry has tried to eliminate past characterizations such as "pre-engineered metal buildings" in lieu of the more accurate identification of "metal building systems".

Metal building systems have evolved over the years into assemblages of structural elements that work together as a very efficient structural system. While there are many variations on the theme, the basic elements of the metal building system are constant: primary rigid frames, secondary members composed of wall girts and roof purlins, cladding, and bracing. Metal building system design may seem trivial at first, but experience shows that the complex interaction of these elements into a stable system is a challenging engineering task. MBMA member companies have demonstrated this expertise and are on the leading edge of systems design. The reality of this has been evident in post-disaster investigations where metal building system "look-a-likes" have suffered more extensive damage than engineered systems subjected to the same extreme loading. These metal buildings are not easily distinguished from a metal building system, however, they are not designed with regard to the interdependence of the structural elements. The individual components are usually purchased from one or more vendors and often shipped out as a complete metal building.

Leon's Carpet & Tile headquarters (above) in Austin, TX, and the Gold Shore Casino (opposite) in Biloxi, MS, demonstrate the versatility of today's metal buildings.

Optimization - The Bottom Line

In an AISC Marketing, Inc. publication (Myths and Realities of Steel Buildings), it is noted that "the cost of a structure is divided among four areas: design, material, fabrication and erection. The optimum structure has the lowest weight consistent with the greatest standardization of parts, the fewest number of pieces and the smallest amount of detail work." Metal building systems offer an excellent optimization solution, taking advantage of lower costs in all of the four areas noted.

Metal building manufacturers have developed design and fabrication techniques to optimize the primary frames and put the material where it is needed. Using welded plate members as opposed to hot-rolled sections provides this opportunity. Optimization is carried out by using tapered webs with increased depth in areas of higher moments, and by varying the web thickness and flange size as needed. Sections are also frequently used with unequal flanges (monosymmetric) to utilize a smaller flange where it might be in tension and/or a larger compression flange in an unbraced condition. The frame members are also designed with bolted end-plate connections for fast assembly in the field. MBMA has sponsored extensive research over the years to develop appropriate design methods for bolted end-plate connections. These connections are receiving new interest as a possible alternative to welded moment connections in seismic areas (for more information, see Ronald L. Meng's and Thomas M. Murray's paper, "Seismic Performance of Bolted End-Plate Moment Connections" in the Proceedings of the AISC National Steel Construction Conference, May 1997).

Cold-formed secondary members (purlins and girts) offer a high strength to weight ratio. Optimized sections can be roll-formed by the metal building system manufacturer to the required depth and thickness to carry the specified loads.

Structural Redundancy

Metal building systems are extremely redundant structures. Structural redundancy is the degree to which multiple load paths exist within a structure.

Shown at top is the Transcypt building in Lincoln, NE, while pictured above is the interior of the Wilkof-Morris plant in Canton, OH.

That is, if a local failure occurs in a member, does the member lose all capacity to carry additional load, or is the load redistributed and carried by the member in a different manner or carried by other members? All flexural members used in a metal building, including panels and purlins, are usually continuous members that permit redistribution of loads as opposed to simply supported members that provide no redundancy. The primary frames are indeterminate structures that offer additional redundancy.

MBMA is sponsoring a study to look into additional design requirements for overload situations, particularly abnormal snow loads. Current codes and standards, such as ASCE 7, require most buildings to be designed using loads that represent a 50-year mean recurrence interval. However, improved performance under abnormal snow load situations may be achieved through specific loading and/or design considerations. MBMA has also sponsored research to determine if the unbalanced snow load specified in most codes adequately reflects the propensity for snow to drift on low slope roofs. Revisions to ASCE 7 are being considered based on the results of this work that should lead to improved performance for all types of buildings with low-slope roofs.

Expansion/Change of Building Use

One of the reasons that metal building systems are an attractive economic alternative, as discussed above, is that materials are optimized for the specified loads. Obviously, if substantial new loads are introduced that were not accounted for in the original design, modifications may be required. An owner actually has three options with regard to a new building design: 1) the building can be overdesigned in anticipation of future potential loads, 2) the building can be economically designed with some consideration for future potential loads, or 3) the building can be designed for the current loads to provide the lowest initial cost. All three of these options are available with a metal building system and would be designed according to the applicable building code, depending on an owner's anticipated needs. The first option should be evaluated against the other two by comparing the initial costs versus modification costs in the future given the probability that the loads will actually be added. In reality, most building types that depend on a limited availability of standard rolled beams and other structural elements are overdesigned by the nature of the selection process (i.e. the next largest available size is used based on the required section). This is where the greater versatility of welded plate frames and cold-formed members provides lower material costs. The second option for future expansion capability is often provided for in a metal building design, by designing the end frame for additional future loads. However, even if this is not pre-planned in the design, using the third option for the lowest initial cost, alterations can readily be made working with the metal building manufacturer.

A session is tentatively planned for the 1998 AISC National Steel Construction Conference to identify how expansions or changes of building use can efficiently be carried out in a metal building system. This should be of interest to engineers involved in such projects, and should help identify the most cost-effective ways to alter a metal building system. This strategy of providing the most economic building to accommodate the specified structural loads, and cost-effective ways to provide for possible future expansion, should be compared to the alternative of overdesigning a building for loads that may not ever be realized.

Architectural Involvement

The growth of metal building systems is also attributable to more diverse applications that have evolved due to the increasing awareness by architects of what metal buildings offer. Many different architectural finishes may be utilized to provide the look required for applications such as churches, schools, shopping centers, office buildings, etc., while still offering the cost advantages of the metal building system. These architectural finishes, other than steel cladding, are typically not designed or provided by the metal building manufacturer. The role of the architect/engineer of record is important in ensuring that the design criteria, particularly the lateral drift and deflection of the primary frame and/or secondary supports, is consistent with the chosen wall finish. This is true of any type of construction, and is easily coordinated with the metal building manufacturer.

In recognition of the importance of serviceability criteria, MBMA jointly sponsored with the American Iron and Steel Institute (AISI) and AISC the development of the AISC Guide on Serviceability Design Considerations for Low-Rise Buildings (see AISC Steel Design Guide Series 3, "Serviceability Design Considerations for Low-Rise Buildings"). This reference establishes guidelines for many serviceability issues, including frame-supported cladding. For example, lateral drift of frames may range from H/60 for flexible metal cladding to H/200 for more brittle reinforced masonry walls. Additionally, MBMA has worked with the National Concrete Masonry Association in the development of its manual to address serviceability as well as structural design considerations when utilizing concrete masonry walls with metal buildings. The availability of these tools has helped give architects more confidence in blending different architectural finishes such as precast concrete, brick, stone, wood and glass with metal building systems. However, these wall finishes are not just used for aesthetic enhancements. For example, in many industrial or warehousing operations, masonry wainscoting can be employed to improve resistance to impact loading from vehicular traffic.

With regard to the trend away from "conventional" construction toward metal building systems, as reflected in the market share analysis, more engineering opportunities are anticipated for those working in this low-rise segment. An increasing number of applications will have an engineer of record or architect involved with the construction project. This design professional assists in the preparation of the design specifications for the project, including the metal building system and its erection, and where appropriate, assists in supervising the construction process for compliance with the contract documents. The metal building manufacturer is responsible only for the structural design of the metal building system it sells to the end customer and for supplying adequate evidence of compliance with the specifications, design criteria, and design loads to the design professional to incorporate the metal building system into the construction project. An engineer of record is not only involved in this incorporation of the metal building system into the project, but also typically has the design responsibility for the foundations, floor slabs, interior and exterior concrete masonry walls, or tilt-up walls, and the connection of these walls to the metal building framing. Engineers and architects who are interested in this growing market should consider this as an excellent opportunity to expand their professional services. MBMA will be sponsoring seminars and short courses in the near future to help inform interested design professionals on how they can better understand and interact with the metal building systems industry.

Standing Seam Roofs

The introduction and use of the standing seam roof by the metal building industry has provided one of the most efficient structural, weathertight systems available today. Over 2 billion square feet of standing seam roofing is installed annually. In fact, more than 50 percent of all commercial/industrial buildings are covered with metal. MBMA has worked with AISI over the last 25 years in sponsoring research to develop design and test procedures for standing seam roof systems. This research has provided new requirements for designing and testing purlins supporting standing seam roofs in the latest AISI Cold-Formed Specification. The bracing of purlins supporting standing seam roofs can be handled in several ways, and is one of the most important considerations for maintaining the structural stability of the system. In recognition of this significance, MBMA has worked with AISI to produce "A Guide for Designing With Standing Seam Roof Panels, CF97-1", which addresses the design considerations for members supporting a standing seam roof.

Standing seam roofs are also providing new life for many buildings with flat roofs that have proved to be problematic. A standing seam roof can be designed to be applied directly over an existing flat, built-up roof using a substructure to provide a low-slope replacement, or a steeper slope for a new architectural appearance. The new standing seam roof will not only provide a superior weathertight covering, but life cycle cost analyses usually show them to be the best economic choice when considering lower costs involved with removing the old roof, minimal disruption of facility operations, and energy-saving capabilities of new insulation in the created "attic" space.

Durability

Metal building systems are extremely durable and require only minimal maintenance. The engineered coating systems that are available provide a long service life, even in extremely corrosive environments. The use of zinc, aluminum and aluminum/zinc alloy coatings has helped increase the life expectancy of steel cladding. In fact, a recent survey of 82 Galvalume roofs up to 22 years old located in the industrial North and along the South Gulf and Atlantic Coasts found them in excellent condition. Based on their appearance, they were projected to last 30 years or longer in most environments before requiring major maintenance. This survey supports the expectations that are the basis for the warranties typically supplied for paint coating performance. While the aluminum/zinc alloy coatings provide the corrosion protection, various types of paint coatings provide the architect with a veritable palette of cladding colors. These coatings include specialized painted fluorocarbons and siliconized polyester systems that perform extremely well, as evidenced by the warranties normally supplied to cover color retention, fading, and chalking for a specified period of time.

AISC Certification Program

As previously mentioned, MBMA will require AISC Certification as a requirement for membership beginning in the year 2000. This is a major endorsement of the quality that is expected of MBMA members and inherent in the AISC Certification program. The program ensures that a metal building manufacturer has achieved a high level of competency in all aspects of design and fabrication. The purpose of the program as stated by AISC is to "confirm to the construction industry that a Certified structural steel fabricating plant has the personnel, organization, experience, procedures, knowledge, equipment, capability and commitment to produce fabricated steel of the required quality for a given category of structural steel work."

The certification program examines policies and procedures at each of the manufacturer's facilities, and the application of those policies and procedures to randomly selected projects. It consists of nearly 200 questions covering the areas of general management, engineering and drafting, procurement, manufacturing, and quality control. Inspection and evaluation teams from an independent auditing firm annually observe and evaluate the manufacturer in almost every aspect of operation. In addition, unannounced and unscheduled inspections verify continuous compliance with the detailed certification standards established by AISC, which has been setting steel construction standards and writing specifications for more than 75 years. As a result of all of these measures, architects, design professionals, building code officials and the insurance industry can be assured that certified metal building systems manufacturers are capable of meeting the industry's highest standards for product and design integrity.

Summary

Involvement in technical activities has been a hallmark of MBMA over the years. MBMA has helped advance the industry by taking the lead in sponsoring millions of dollars of research conducted by many eminent investigators at prestigious universities. This research has elevated the state-of-the-art in such areas as wind loads on low-rise buildings, tapered member analysis and design, bolted end-plate connections, and cold-formed steel design. These efforts have led to technically advanced systems that are some of the best low-rise construction alternatives available. The efficient use of materials is only surpassed by the performance of the metal buildings and the satisfaction of hundreds of thousands of building owners.

Continued advancements are planned to help take advantage of new design methodologies, new material advances, and new fabrication and construction techniques. The metal building systems of today bear little resemblance to the Quonset Huts or "tin-shed" ancestors that many still associate with the industry. This is in large part due to the progressive approach that has helped metal building systems remain on the leading edge of low-rise construction.

W. Lee Shoemaker, Ph.D., P.E. is Director of Research & Engineering with the Metal Building Manufacturers Association in Cleveland.

Appendix E Framing Systems

AN INTEGRATED DESIGN APPROACH OFFERS FLEXIBILITY, ECONOMY, DURABILITY
An Evaluation of Metal Building Systems

METAL BUILDING SYSTEMS ELEMENTS

Metal building systems offer a completely integrated set of interdependent elements and assemblies, which, taken together, form the total building. Included are primary and secondary framing elements, covering components, and accessories. These building block parts come in many different configurations, as illustrated below.

Framing Systems

Rigid frame clear span: Provides a column-free interior space 20 feet to 160 feet and wider. Recommended applications include auditoriums, gymnasiums, warehouses and aircraft hangars.

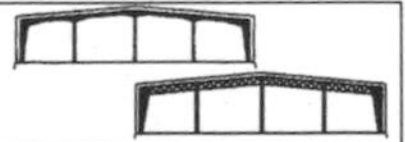

Ridge frame multi-span (solid web rafter) and modular open web (open web rafter) — 50 feet to 500 feet and wider. Used where interior columns do not impair the function of the building. Interior columns shorten the spans of the rafter beam, thereby often reducing the frame cost. Recommended applications include manufacturing facilities and warehouses.

Flush wall clear span—20 feet to 70 feet and wider. Offers not only column-free interior floor space but also uniform depth columns. The secondary wall structural systems (girts) are totally flush with the interior flanges of the columns. This allows interior sidewalls to be finished without the frame columns protruding into the interior wall line. Recommended applications include retail stores, branch banks, and office facilities.

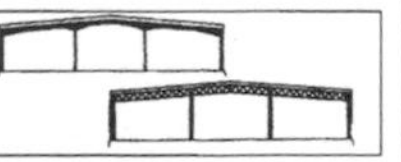

Flush wall multi-span—50 feet to 250 feet and wider. Available with solid or open web rafters. Uniform depth exterior columns with girts are totally flush with the inside column flange. Interior columns reduce span lengths, thereby reducing costs. Recommended applications are buildings where interior sidewalls are to be finished for office areas or for warehouses and distribution centers where "close to wall' palletizing is required.

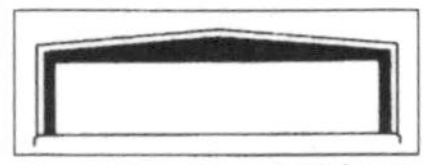

Tapered beam straight columns—15 feet to 70 feet and wider. Clear span with uniform depth columns. Greater vertical and horizontal clearance at column/rafter connection than rigid frame clear span. This system is economical for narrow widths. Recommended applications include offices, retail stores, and buildings with bridge crane systems.

Single slope clear span—20 feet to 160 feet and wider. Available with solid or open web rafters. This system provides single direction roof slope for rainwater runoff control. It is often used for shopping centers and office complexes where rainwater must be directed away from parking areas.

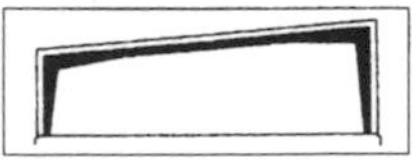

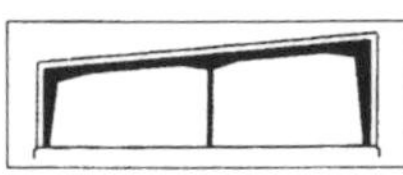

Single slope multi-span—50 feet to 200 feet and wider. Available with solid or open web rafters. The single slope design facilitates rainwater runoff control. This system is used in facilities where interior columns do not impair building function. Recommended applications—manufacturing, warehousing, and distribution centers, retail shopping centers, and office complexes.

Lean-to—10 feet to 60 feet and wider. Used primarily for wing units and additions to existing facilities. A lean-to can be used with any of the above framing types.

Index

About the Author

William D. Booth has more than 37 years' experience in building materials sales, metal building sales, and running his own company specializing in metal building sales and construction. He is the author of three other books, including one on boating published by Cornell Maritime Press, *Marketing Strategies for Design-Build Contracting*, from Chapman & Hall, and *Selling Commercial and Industrial Construction Projects*, published by Van Nostrand Reinhold